SEM Micrograph of the complex trabeculated inner side of the apex in rat right ventricle from Endocardial Endothelium: Functional Morphology by Luc J. Andries

MEDICAL
INTELLIGENCE
UNIT

ENDOCARDIAL ENDOTHELIUM: CONTROL OF CARDIAC PERFORMANCE

Stanislas U. Sys, M.D., Ph.D.
Dirk L. Brutsaert, M.D., Ph.D.

University of Antwerp
Antwerp, Belgium

Springer
New York Berlin Heidelberg London Paris
Tokyo Hong Kong Barcelona Budapest

R.G. LANDES COMPANY
AUSTIN

Medical Intelligence Unit

ENDOCARDIAL ENDOTHELIUM: CONTROL OF CARDIAC PERFORMANCE

R.G. LANDES COMPANY
Austin, Texas, U.S.A.

Submitted: February 1995
Published: July 1995

Please address all inquiries to the Publisher:
R.G. Landes Company, 909 Pine Street, Georgetown, Texas, U.S.A. 78626
or
P.O. Box 4858, Austin, Texas, U.S.A. 78765
Phone: 512/ 863 7762; FAX: 512/ 863 0081

U.S. and Canada ISBN 1-57059-236-5

International ISBN 3-540-59431-0

Library of Congress Cataloging-in-Publication Data

Endocardial endothelium: control of cardiac performance / edited by
Stanislas U. Sys and D. L. Brutsaert.
p. cm. -- (Medical intelligence unit)
Includes bibliographical references and index.
ISBN 1-57059-236-5
1. Endocardium--Physiology. 2. Heart--Physiology. I. Sys,
Stanislas U., 1954- . II. Brutsaert, D. L. (Dirk L.)
III. Series.
[DNLM: 1. Endocardium--physiology. 2. Cardiac Output--physiology.
WG 285 E557 1995]
QP114.E53E53 1995 612.1'7--dc20
DNLM/DLC 95-16984
for Library of Congress CIP

Publisher's Note

R.G. Landes Company publishes five book series: *Medical Intelligence Unit, Molecular Biology Intelligence Unit, Neuroscience Intelligence Unit, Tissue Engineering Intelligence Unit* and *Biotechnology Intelligence Unit.* The authors of our books are acknowledged leaders in their fields and the topics are unique. Almost without exception, no other similar books exist on these topics.

Our goal is to publish books in important and rapidly changing areas of medicine for sophisticated researchers and clinicians. To achieve this goal, we have accelerated our publishing program to conform to the fast pace in which information grows in biomedical science. Most of our books are published within 90 to 120 days of receipt of the manuscript. We would like to thank our readers for their continuing interest and welcome any comments or suggestions they may have for future books.

Deborah Muir Molsberry
Publications Director
R.G. Landes Company

CONTENTS

EDITORS

Stanislas U. Sys, M.D., Ph.D.
Professor of Physiology
Department of Physiology
University of Antwerp
Antwerp, Belgium
Chapters 1, 3, 5, 6

Dirk L. Brutsaert, M.D., Ph.D.
Professor of Physiology and Medicine
Department of Physiology and Medicine
University of Antwerp
Antwerp, Belgium
Chapters 1, 6

CONTRIBUTORS

Luc J Andries, Ph.D.
Department of Physiology
University of Antwerp
Antwerp, Belgium
Chapters 1, 2, 6

Gilles W. De Keulenaer, M.D.
Department of Physiology
University of Antwerp
Antwerp, Belgium
Chapters 3, 5, 6

Marc Demolder
Department of Physiology
University of Antwerp
Antwerp, Belgium
Chapter 4

Paul Fransen, Ph.D.
Department of Physiology
University of Antwerp
Antwerp, Belgium
Chapter 4

Grzegorz Kaluza, M.D.
Department of Physiology
University of Antwerp
Antwerp, Belgium
Chapter 6

Puneet Mohan, M.D.
Department of Physiology
University of Antwerp
Antwerp, Belgium
Chapter 5

FOREWORD

This book is about the key role of cardiac endothelial cells in the control and modulation of the performance of the heart. In 1986, Brutsaert and collaborators made the novel and intriguing observation that selective damage of endocardial endothelial cells lining the inner surface of the cardiac cavities resulted in a profound alteration of the contractile performance of the immediate subjacent cardiomyocytes. Subsequent investigations have led to similar observations on the direct modulatory effect of cardiac microvascular endothelial cells on subjacent cardiomyocytes.

The book provides a state of the art of this novel concept placing it in a historical perspective and emphasizing how, through the contribution of numerous investigators, it progressed towards full maturity.

The present book on "Endocardial Endothelium: Control of Cardiac Performance" by S.U. Sys and D.L. Brutsaert logically follows publication of a previous monograph on "Endocardial Endothelium: Functional Morphology" by L.J. Andries from the same laboratory at the University of Antwerp.

PREFACE

This book on the physiopharmacological aspects of the endocardial endothelium logically follows publication of a previous book in the same monograph series by R.G. Landes Company on the functional morphological aspects of the endocardial endothelium.[1] In an introductory chapter 1, it is summarized how the novel concept of the obligatory role of cardiac endothelial cells in the control and modulation of cardiac performance was developed between 1983 and 1986 by Brutsaert and collaborators at the University of Antwerp, and how thereafter it slowly progressed to full maturity through the contribution of many investigators. The authors wish to acknowledge the numerous past and present collaborators in the laboratory of physiology at the University of Antwerp who have contributed in elaborating this concept: Andries L, De Hert S, De Keulenaer G, De Mey J, Demolder M, Fransen P, Gillebert T, Housmans P, Jagenau T, Kaluza G, Leite-Moreira A, Meulemans A, Mohan P, Rouleau J, Schoemaker I, Shah A, Sipido K, Vandenbroucke M. Chapter 2 summarizes the major morphological, embryological and comparative physiological features of cardiac endothelial cells. In chapter 3, the experimental observations on how cardiac endothelial cells affect the mechanical performance of the heart are described.

As for the underlying mechanisms of the interaction between cardiac endothelial cells and cardiomyocytes, two working hypotheses have been postulated over the past years, i.e. (i) interaction mediated through a trans-endothelial physico-chemical gradient for various ions (active blood - heart barrier), and (ii) interaction mediated through the release by the cardiac endothelial cells of various cardio-active substances, as e.g. nitric oxide (NO), endothelin (Et) and prostacyclin (PGI_2). These two mechanisms which may act in concert or in parallel have been elaborated in chapters 4 and 5 respectively. Finally in chapter 6, it is speculated how dysfunctional cardiac endothelial cells could contribute to the pathogenesis of various cardiac diseases.

1. Andries LJ, Endocardial Endothelium: Functional Morphology. R.G. Landes Company, Biomedical Publishers Austin (USA). pp. 134, ©1994.

BIOGRAPHIES

Stanislas U. Sys

Stanislas U. Sys (born in Lo, Belgium, 1954) is full professor in the Department of Physiology and Medicine at the University of Antwerp (Belgium) since 1992, teaching courses in Human Organ Physiology and Pathophysiology and in Medical Statistics. He earned his Ph.D. in Mathematics at the University of Ghent (Belgium) and his MD and Ph.D. degrees at the University of Antwerp (Belgium). He is the author or co-author of 38 full peer-reviewed papers in cardiovascular physiopharmacology. His scientific interest mainly focused on the mechanical function of the heart as a muscle and pump, where he contributed to problems related to myocardial relaxation, to ventricular relaxation and filling as well as to diastole, to the importance of functional non-uniformities during contraction and relaxation in the heart, and more recently to the obligatory role of the cardiac endothelial cell in the control and modulation of myocardial and cardiac performance. As a mathematician, his interest also focused on quantification and biomathematical modeling of myocardial and cardiac performance and on study design, data management and interpretation in medical research.

Dirk L. Brutsaert

Dirk L. Brutsaert (born in Ghent, Belgium, 1937) is full professor in the Department of Physiology and Medicine at the University of Antwerp (Belgium) since 1966, teaching courses on Human Organ Physiology and Pathophysiology. He earned his MD and Ph.D. degrees at the University of Ghent (Belgium). As a board certified cardiologist, he has, in addition, a joint appointment as adjunct director of cardiology at the University Hospital of the University of Antwerp. He has been Visiting Lecturer at Harvard Medical School (1968-1972), Visiting Professor at the University of California, San Francisco (1976, 1979), and at McGill University (1987). He is the author or co-author of more than 170 full peer-reviewed papers in Cardiovascular Physiopharmacology and Cardiology. His scientific interest mainly focused on the mechanical function of the heart as a muscle and pump, where he contributed to problems related to cardiac contraction, such as the concept of contractility, to ventricular relaxation and filling as well as to diastole, to the importance of functional non-uniformities in the heart, and more recently to the obligatory role of the cardiac endothelial cell in the control and modulation of myocardial and cardiac performance.

CHAPTER 1

INTRODUCTION

Dirk L. Brutsaert, Stanislas U. Sys and Luc J. Andries

Following the observations by Furchgott & Zawadski[9] about the obligatory role of vascular endothelium in vasomotor tone, we wondered whether endothelial cells in the heart — either endocardial endothelial cells or cardiac microvascular endothelial cells in the coronary vascular tree (Fig. 1.1) — could similarly, but apart from their indirect effect through modulation of the coronary perfusion, play an obligatory role in the direct control of myocardial and cardiac performance through their effects on the immediate subjacent cardiomyocytes.

Preliminary experiments were performed in our laboratory between 1983 and 1985. These initial experiments have led — in 1986[4-6] — to the intriguing but convincing observation that selective damage of endocardial endothelium in isolated cardiac muscle, i.e. papillary muscle from cat and rat, resulted in a typical response. Following damage, the isometric twitch was shorter in duration by an earlier onset of isometric force decline. This twitch abbreviation was accompanied by a slight decrease in peak twitch force development but with no significant changes in the rate of rise of the twitch.[7] Hence, it seemed that, conversely, an intact endocardial endothelium enhanced, or at least sustained myocardial performance by prolonging twitch duration thereby increasing peak performance.[1]

This fundamental observation has subsequently been confirmed by numerous investigators (for references see chapters 2 through 5) from other laboratories, in cardiac muscles from different animal

Endocardial Endothelium: Control of Cardiac Performance, edited by Stanislas U. Sys and Dirk L. Brutsaert.

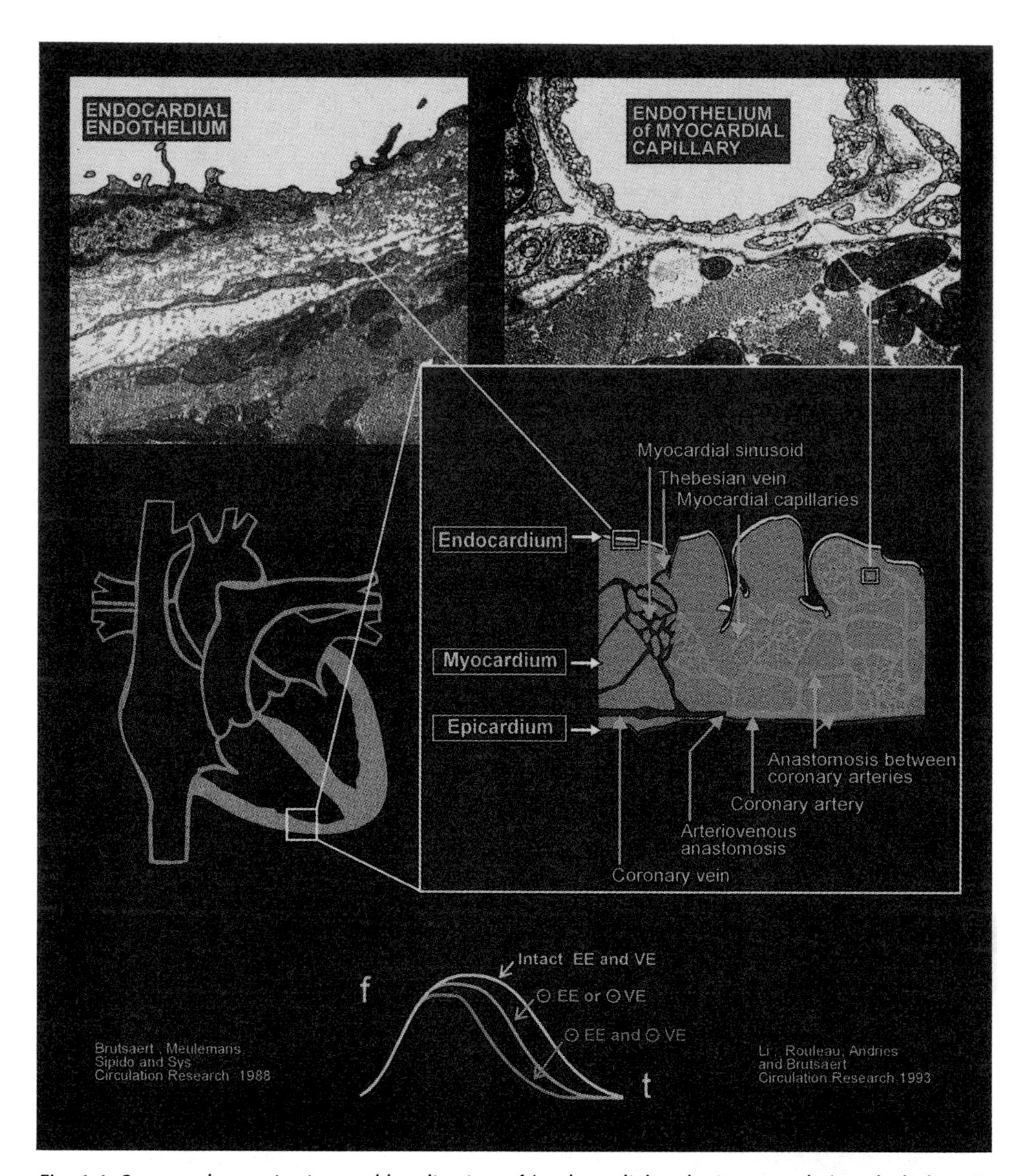

Fig. 1.1. Structural organization and localization of (endocardial and microvascular) endothelium in the heart. The diagram shows part of the ventricular wall. This structure is delineated by: the epicardium, *a smooth outer layer of mesenchymal cells and interstitial tissue, and the* endocardium, *which includes a luminal layer of endothelial cells, the* endocardial endothelium *(**EE**), and the underlying basement membrane and fibroelastic layer (see chapter 2). The luminal surface has a complex architecture with numerous fissurae, sinuses, papillary muscles and trabeculae carneae, resulting in a high surface-to-volume ratio, and hence a large contact surface between endocardial endothelium and the circulating blood. The* myocardium *contains a complex system of coronary blood vessels and sinusoids which are all lined by an intimal layer of* vascular endothelial (**VE**) cells. *In the myocardium, anastomoses occur between coronary arteries and between coronary arteries and veins. Other particular connections include the Thebesian veins, which drain blood from the coronary veins and sinusoids towards the ventricular lumen, and the arterioluminal vessels (not shown) which form communications between coronary arteries and the ventricular cavity. Through this internet of communications the endocardial and microvascular endothelia form one contiguous structure.*

The two transmission electron micrographs show a detail of the EE (upper left) and VE of a myocardial capillary (upper right). The luminal surface of the endocardium consists of a thin layer of closely apposed EE cells, which are in contiguity with VE cells of the aorta and other macrovessels. The thickness of the interstitial tissue between EE and subjacent myocytes ranges from 1 μm in small mammals to more than 50 μm in humans. The endocardium is much thicker in atria than in the ventricles. The capillary-myocyte distance is usually less than 1 μm. Endothelium of myocardial capillaries, like EE, has been classified as a continuous type of endothelium. However, considerable morphological differences exist between both types of endothelium, as is further discussed in chapter 2.

*The effects of the presence of intact EE and VE, and the absence of one or both endothelial types is summarized in the lower graph (bottom). The curves represent control (***Intact EE and VE***) and experimental isometric force (f) twitches from isolated papillary muscles in Krebs-Ringer solution with 1.25 mM Ca^{2+}. In one group of experiments, EE of isolated papillary muscles was selectively removed by a quick immersion with Triton X-100 (***- EE***). In a second group of experiments, VE in Langendorff hearts was first damaged or made dysfunctional by a bolus injection of Triton X-100 in the coronary arteries. The EE in these hearts was still intact. From the Triton-treated Langendorff hearts, papillary muscles (***- VE***) were isolated and their contractile characteristics were compared with control muscles. Finally, the EE of the -VE muscles was damaged by quick immersion in Triton X-100 (***- EE and - VE***). The summarizing graphs illustrate that removal of VE has similar effects as removal of EE. Both endothelial types have an additive or complementary contraction-prolonging effect on myocardium. Removal of EE from -VE muscles resulted in further shortening of twitch duration and a further decrease of total tension.*

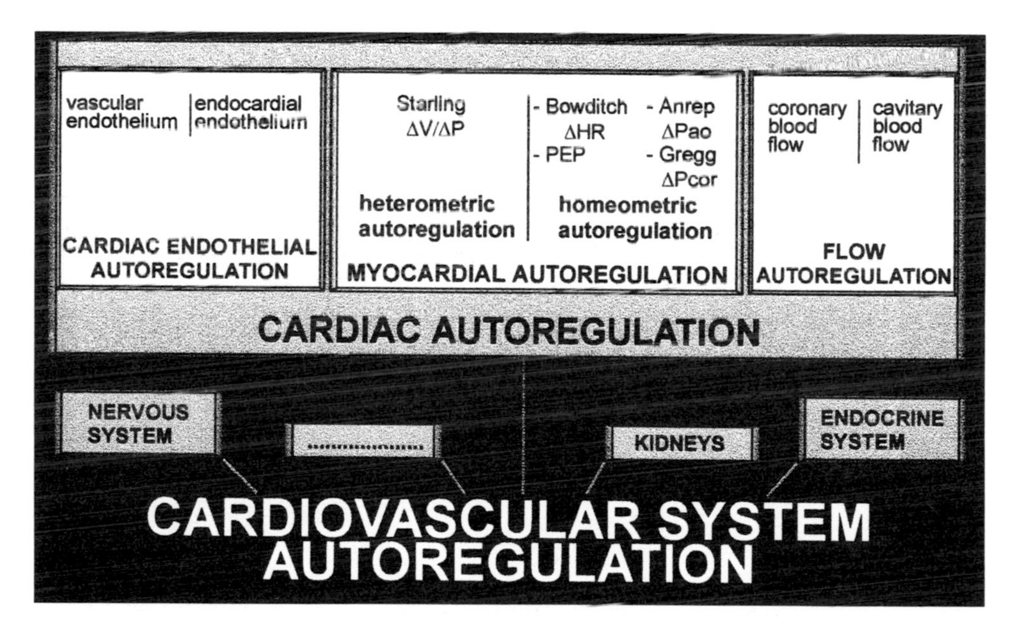

Fig. 1.2. Integrated configuration of myocardial, cardiac and cardiovascular system autoregulation. Cardiovascular performance relies largely on autoregulatory mechanisms including modulation by various other systems in the body. At the myocardial level, heterometric autoregulation is accomplished through Starling mechanisms (ΔV or ΔP: end-diastolic volume or pressure changes) and homeometric autoregulation through changes in the pattern of stimulation (Bowditch ΔHR:changes in heart rate or PEP: post extrasystolic potentiation) and through changes in aortic pressure (Anrep ΔPao) or coronary perfusion pressure (Gregg ΔPcor). The endocardial and (micro)vascular endothelium contribute to the intermediate level of cardiac autoregulation through a direct and sustained inotropic action on the myocardium (cardiac endothelial autoregulation), through modulation or exclusive mediation of the myocardial effect of substances in the (su)perfusing blood and through direct sensitivity to flow of cavitary or coronary blood (flow autoregulation).

species, and also in the in vivo intact animal.[8,10] From these observations we have postulated the existence of an endocardium-mediated intracavitary autoregulation of cardiac performance (Fig. 1.2).[2,3] Under autoregulation of cardiac performance, we understand that the heart as an open regulatory system incorporates all components, including afferent and efferent pathways, to feedback some aspect(s) of its performance as a muscular pump.

In addition, more recent studies have shown that the vascular endothelial cells in the cardiac micro-vasculature, similar to the endocardium-mediated control of subjacent cardiomyocytes, also directly affect contractile performance of the immediate subjacent cardiomyocytes.[11] This would imply that, in addition to the above postulated endocardium-mediated intracavitary autoregulation of cardiac performance, there may be an intracoronary (micro)vascular endothelium-mediated autoregulation of cardiac performance as well.

Meanwhile, the concept of a direct cardiac endothelial control of myocardial and cardiac performance has progressed to an advanced state of maturity through additional original contributions of many researchers from other laboratories worldwide. Their papers are cited in the respective chapters of this monograph.

In conclusion, it would seem that all cardiac endothelial cells, regardless of whether they are from endocardial or from (micro)vascular origin, probably directly control or modulate the contractile state of the subjacent cardiomyocytes. Autoregulation by both types of endothelial cells acts through alterations in the duration of systole by modulating the onset of ventricular relaxation and rapid filling of the heart.

REFERENCES

1. Brutsaert DL. The endocardium. Annu Rev Physiol 1989; 51:263-273.
2. Brutsaert DL. Endocardial and coronary endothelial control of cardiac performance. NIPS 1993; 8:82-86.
3. Brutsaert DL and Andries LJ. The endocardial endothelium. Am J Physiol 1992; 263:H985-H1002.
4. Brutsaert DL, Meulemans AL, Sipido KR et al. Mechanical performance of the myocardium is modulated by the endocardium. Biophys J 1987; 51:464a.
5. Brutsaert DL, Meulemans AL, Sipido KR et al. Endocardial control of myocardial performance. In: Sugi H and Pollack GH, ed.

Molecular mechanism of muscle contraction. Proceed. of the 1986 Muscle Symposium, Hakone Japan: Plenum Publ, 1986:609-616.
6. Brutsaert DL, Meulemans AL, Sipido KR et al. The endocardium modulates the performance of myocardium. Arch Int Physiol Biochem 1987; Louvain, Nov. 1986; 95:4.
7. Brutsaert DL, Meulemans AL, Sipido KR et al. Effects of damaging the endocardial surface on the mechanical performance of isolated cardiac muscle. Circ Res 1988; 62:357-366.
8. De Hert SG, Gillebert TC and Brutsaert DL. Alteration of left ventricular endocardial function by intracavitary high-power ultrasound interacts with volume, inotropic state, and a_1-adrenergic stimulation. Circulation 1993; 87:1275-85.
9. Furchgott RF and Zawadski JV. The obligatory role of endothelial cells in the relaxation of arterial smooth muscle by acetylcholine. Nature 1980; 288:373-376.
10. Gillebert TC, De Hert SG, Andries LJ et al. Intracavitary ultrasound impairs left ventricular performance: presumed role of endocardial endothelium. Am J Physiol 1992; 263:H857-H865.
11. Li K, Rouleau JL, Andries LJ et al. Effect of dysfunctional vascular endothelium on myocardial performance in isolated papillary muscles. Circ Res 1993; 72:768-777.

CHAPTER 2

Functional Morphology

L.J. Andries

Endothelial cells are highly active and multifunctional cells which are intensively involved in intercellular communication, in controlling blood coagulation and platelet aggregation, and in the regulation of transendothelial transport. Endothelial cells constitute a heterogeneous population with distinct morphological and physiological properties in the different segments of the vascular tree and in vessels of different organs. The *endocardial endothelium* (EE) forms the inner lining of a unique part of the cardiovascular system, through which all circulating blood passes. The EE is subjected to considerable physico-mechanical stress by the large variations in shape of the heart walls during the cardiac cycle and by the large differences in hydrostatic pressure. These unique features probably have consequences for structure and physiology of the EE.

1. STRUCTURE OF THE ENDOCARDIUM

In vertebrates, the luminal side of the heart has a complex structure consisting of fissurae, cylinder- and sheet-like trabeculae and papillary muscles. In the human heart, trabeculations are usually more complex in the right than in the left ventricle.[1,13] The complex cavitary surface of the ventricular wall is completely covered by the endocardium. The latter structure is visible as a thin transparent to white tissue layer covering the myocardium. It comprises a thin luminal layer of cells, the EE and subjacent, a basement membrane and a fibro-elastic layer.[79] EE cells are thin with bulg-

Endocardial Endothelium: Control of Cardiac Performance, edited by Stanislas U. Sys and Dirk L. Brutsaert.

ing nuclei and a variable number of microvilli.[15,33,86] EE cells usually have a polygonal cell shape[5,33] and are closely apposed to one another with well-developed intercellular borders[6,7,63] The immediate underlying basal membrane consists of a basal lamina and a reticular lamina with fine collagen fibers which merge with fibers from the fibroelastic layer. The fibroelastic layer is a matrix formed by collagen and elastic fibers and it contains fibroblasts, myofibroblasts and smooth muscle cells, which often forms a discontinuous layer as, e.g., in atria.[48] In addition, a heterogeneous population of nerve fibers is present in the fibroelastic layer[56] consisting of sensory nerve subpopulations and presumed parasympathetic, acetylcholinesterase positive nerves. Varicose nerve fibers were found in the proximity (0.2 µm to 0.4 µm) of overlying EE cells.[56]

The elastic fibers in the fibroelastic layer increase in size and number near the subendocardium.[79] There is no distinct border between the fibroelastic layer and the subendocardium. The presence of thick collagen fibers and capillaries is characteristic for the subendocardium[48,79] which is continuous with the perimysium of the myocardium. Other authors have used a slightly different terminology for the endocardial layers and for the border zone between endocardium and myocardium (Fig. 2.1).[22,48,52,63] The fibroelastic layer of the endocardium in the human heart is variable in thickness. The endocardium is thin in both ventricles, ranging from 50 to 300 µm.[59] In atria, especially the left atrium, the endocardium is much thicker and ranges from 600 to 900 µm.[59]

2. COMPARATIVE MORPHOLOGY

In small mammals and particularly in lower vertebrates, the endocardium forms a very thin layer. In frog[87,105] and fish[53,112] hearts, only a thin layer of interstitial tissue of less than 1 µm thickness separates the continuous EE from underlying myocytes. The invertebrates in general have a simple open cardiovascular system devoid of endothelial cells (for a review see Andries[3]). Only mollusks such as the very active Cephalopoda have a well developed and nearly closed cardiovascular system[23,51] with a discontinuous EE and vascular endothelium. Phylogenetically, the first continuous EE becomes apparent in amphioxus[23] and in the hagfish (Agnatha).[131] A remarkable feature in lower vertebrates is the presence of coronary vessels, which is more related to ecophysiological factors than to the fact that these animals have reached a higher

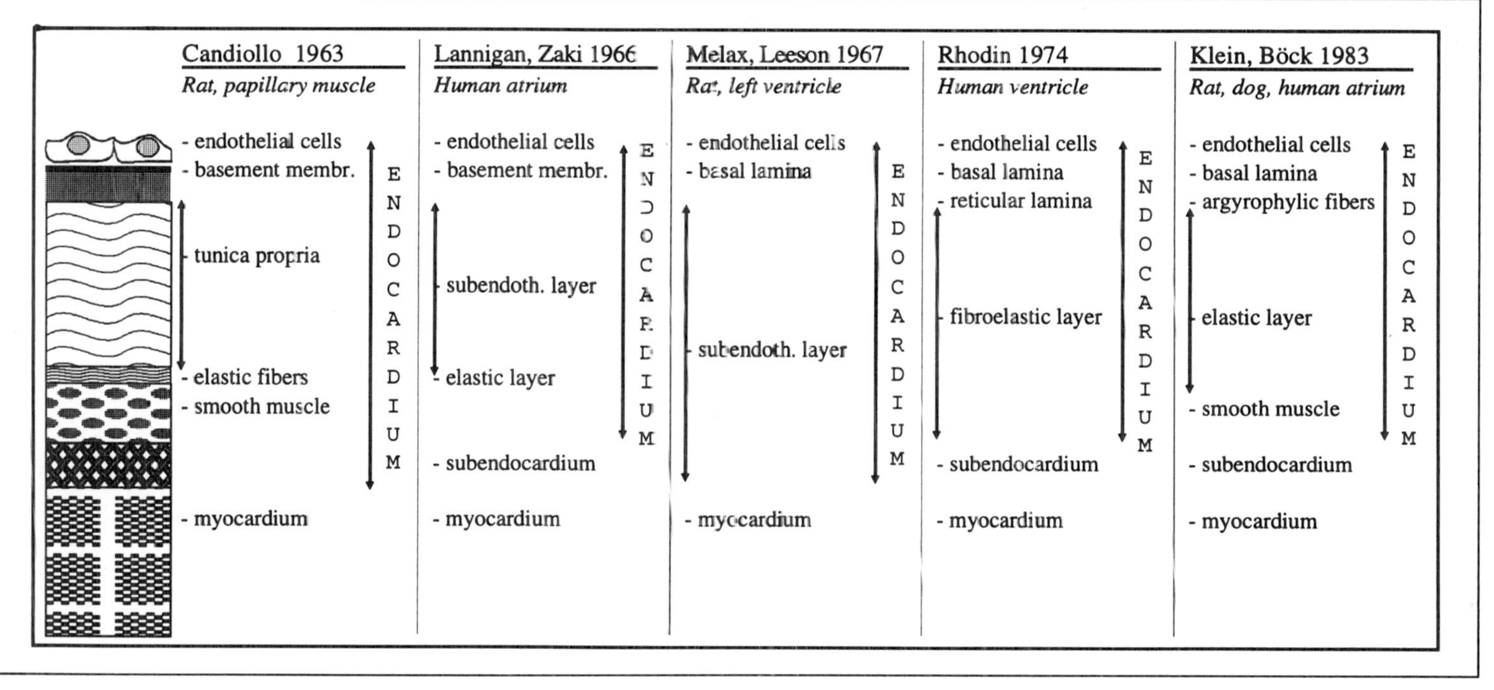

Fig. 2.1. Time chart of the development of endocardial and vascular endothelium in avian (chick) and mammalian (human) embryos. (A) early development; (B) late development. Chick embryo is a widely used model for the study of early development in vertebrates because avian and mammalian hearts are similar at early stages and access to embryo is relatively easy. HH, Hamburger Hamilton; EE, endocardial cells. Reprinted with permission from Brutsaert DL, Andries LJ, Am J Physiol 1992; 263:H985-H1002.

phylogenetic stage. For example, in some sharks like the rays (Elasmobranchs) the coronary system is more complex than in many higher developed fishes and amphibians.[41] In fish, two main types of ventricular organization have been specified:[2,71] (a) a spongy myocardium (*spongiosa*) which is nourished by lacunae between the trabeculae, coronary vessels being completely absent; (b) a spongy myocardium surrounded by an outer compact myocardium (*compacta*) with coronary vessels (Table 2.1). Although coronary vessels are generally restricted to the compacta, they can also vascularize the spongiosa in some species.[2,112,113] The spongious type of ventricular organization is found in "nonathletic" fish like the Japanese Medaka (Oryzias latipes),[53] Scorpaena and Pleuronectes.[113] In these hearts, the EE appears to be metabolically very active[53] and is the only type of endothelium in the heart (Table 2.1). In more athletic and very active fish, like tuna, but also in sharks, the outer compact myocardium contains coronary vessels which reach into the spongy myocardium where they can form numerous Thebesian-like shunts.[2,113] Remarkably, even in mammals, connections exist between the coronary system and the intertrabecular system, the so-called Thebesian vessels.[41] These latter vessels have been suggested to play an important role in the right ventricle and in the interventricular septum. The ventricular wall in small

Table 2.1. Different levels of cardiac organization in fishes

	I. Spongy	**II. Mixed**	**III-IV. Mixed**
Ventricular myoarchitecture	spongiosa	spongiosa compacta	spongiosa compacta
Degree of vascularization	absent or limited to epicardium	limited to compacta	vessels in spongiosa and compacta
Interface	EE	EE + VE	EE + VE
Representative species	scorpion fish, icefish	conger eels, salmon species	sharks, torpedo, tuna,

Fishes of type II to IV all have ventricles consisting of a spongiosa and compacta. The distinction is based on the degree of vascularization and the relative volume of the compacta. The compacta in type III is thinner (<30%; torpedo) than in type IV (>30%;. sharks, tuna). Both types have coronary vessels in compacta and spongiosa and were therefore placed in the same column. The interface between blood and myocardium is formed by EE (endocardial endothelium) in type I, EE and VE (vascular endothelium) in type II to IV. Modified after Tota et al.[2,112]

amphibia, such as *Rana temporaria* and *Rana ribibunda* is merely of the spongy type without a coronary circulation.[71] Larger frogs may have an outer compact myocardium with epicardial coronary vessels. The ventricles of reptilia also consist of an internal spongy and an outer compact myocardium.[71] Capillaries can be found both in the spongy and in the compact myocardium. In birds and mammals, the compact layer of the myocardium is much thicker than in cold-blooded animals and the coronary system becomes more important than EE for the supply of nutrients and oxygen.

In mammals during early embryonic life, the myocardium is, like in lower vertebrates, highly spongious without a coronary circulation (for a review see Brutsaert and Andries[15]). From the 2nd day until nearly the 5th week of development (Fig. 2.2), the EE is the only type of endothelium present in the human heart. Among numerous other functions, it appears to be involved in the formation of trabeculae and in development of cushion tissue. Coronary veins and capillaries appear in 5 week old human embryos, whereas coronary arteries appear later still at an age of 7 weeks. EE cells are probably ontogenetically different from endothelial cells of the coronary system. EE cells are indeed formed very early in embryonic life and originate from the cardiogenic plate, whereas endothelial cells of the coronary vessels appear much later and are derived from mesothelial cells in the epicardium. This ontogenetic difference is also reflected in functional and structural heterogeneity of EE and coronary endothelium.

3. INTERCELLULAR COMMUNICATION

3.1. Endocardial Endothelial Receptors and Inotropic Factors

In all mammalian species the inner surface of the right and left ventricular wall is characterized by the presence of deep crevices in the myocardium and by a prominent trabecular structure, being most pronounced near the apex of the heart. The intense network of trabeculae and the numerous microvilli covering the EE cells augment the available contact surface area which separates the EE from the superfusing blood. There is growing evidence that EE possesses a sensory function for various substances circulating in the superfusing blood.[15] These substances could then release inotropic factors from the EE via a *receptor-mediated*

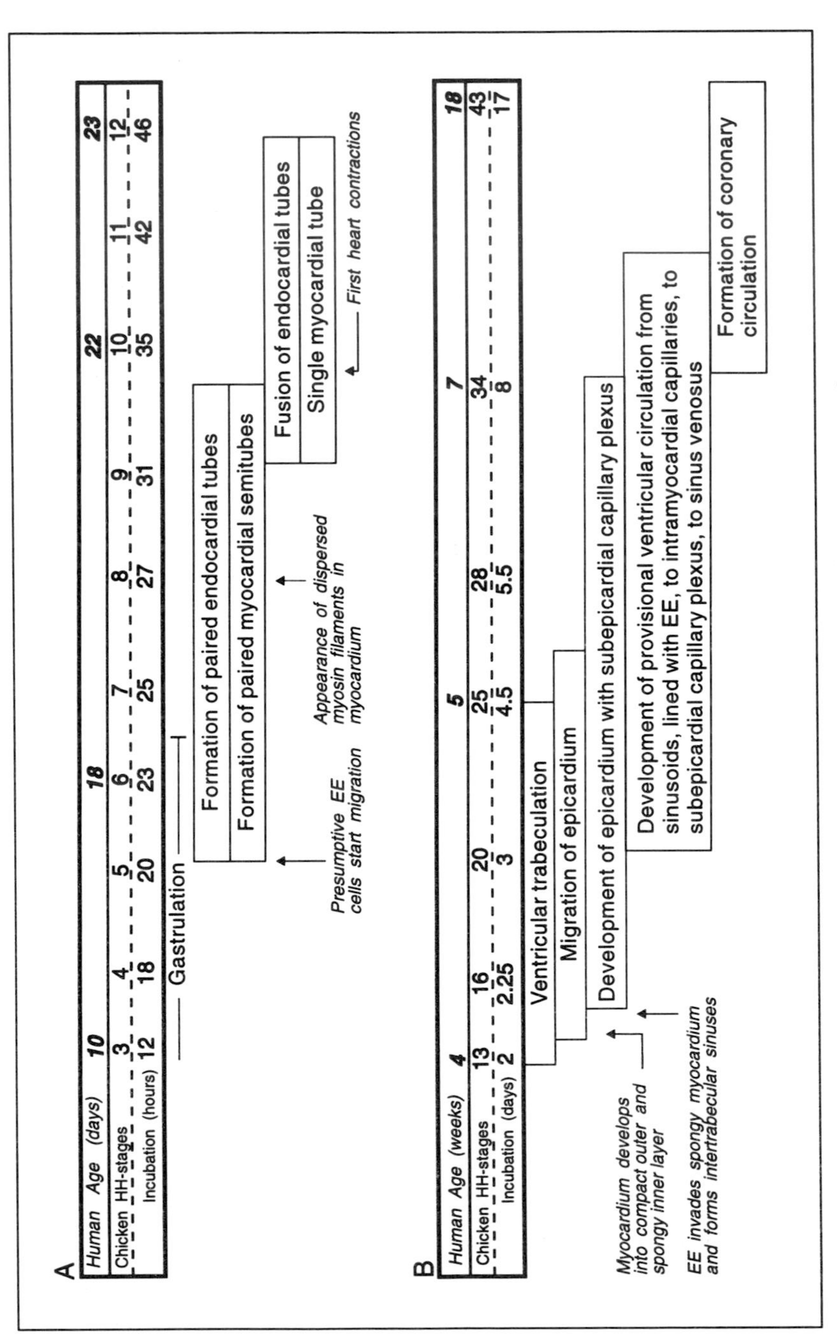

Fig. 2.2. Schematic drawing of the subsequent layers in endocardium according to various authors.

mechanism. It has thus been demonstrated that EE possesses receptors for phenylephrine,[64] vasopressin,[89] angiotensin,[65] serotonin[91] and atrial natriuretic peptide (ANP).[66] ANP receptors consist of homologous cell transmembrane proteins with a cell type-specific differential distribution in primates.[127] In the heart, mRNA for ANP receptor B, a membrane guanylyl cyclase, is abundantly expressed in the EE of right and left atria and of the right ventricle[127] whereas mRNA for ANP receptor C, which is not coupled to cyclic GMP production, is expressed most prominently in EE of all four chambers and throughout the myocardium only in the right ventricle.[127] A nonuniform distribution in cardiac endothelium was also reported for:

- angiotensin converting enzyme which has a low number of binding sites in ventricular EE but high number in valvular EE and coronary vessels; left atrium and both ventricles expressed less binding sites than in the right atrium;[130]
- mineralocorticoid receptors which were present in atrial and ventricular EE and myocytes, but not in small myocardial vessels and capillaries;[54]
- endothelin A and B receptors which were found in EE with the highest density on valvular endothelium, in myocardium and in coronary arteries.[70,124]

The hypothesis that *EE releases inotropic factors*,[16] was substantiated by the finding of diverse bioactive compounds in EE.

- EE cells contain endothelin-1 mRNA[61] and endothelin-1 peptides.[85] Quantitative studies *in vivo* suggest that endothelin levels are higher in EE and endothelium of coronary arteries than in cardiac microvascular endothelium.[42]
- Cultured EE cells release prostaglandins both in basal conditions and after stimulation.[60] Prostacyclin (PGI_2) production was 10 times greater than PGE_2 production. Under flow conditions, PGI_2 release was also much higher in EE cells than in endothelial cells of great vessels.[62]
- Nitric oxide (NO) production in cultured EE cells and in valvular endothelium was determined either by directly measuring NO release or by using bioassay techniques.[90,92,98] In endothelial cells under basal conditions,

NO is processed from L-arginine by constitutive nitric oxide synthase (cNOS). cNOS was detected by the NADPH-diaphorase technique in EE, in endothelium of coronary arteries and veins, but also in endothelium of myocardial capillaries.[115] Immunofluorescent staining confirmed the presence of cNOS in EE and in endothelium of myocardial capillaries (unpublished results).

The markedly nonuniform distribution of various receptors and bioactive compounds further emphasizes the heterogeneity of cardiac endothelium.

3.2. Adhesion Receptors

Expression of endothelial leukocyte adhesion receptors and expression of antigens of the major histocompatibility complex (MHC) play a key role in immunologic reactions. The expression of adhesion receptors constitutes an essential step in the interaction between endothelial cells and leukocytes, in particular in the migration of the latter into underlying tissues during graft rejection, after ischemia and in other cardiac diseases. Interestingly, in cardiac endothelial cells of normal human hearts a distinct heterogeneous distribution appears to exist of adhesion molecules and MHC molecules. ICAM-1 (an adhesion molecule of the immunoglobulin superfamily which interacts with the CD11a/CD18 integrins of leukocytes) is constitutively expressed in endothelium of the entire cardiac system but most intensively in capillaries (Table 2.2).[72,110] In contrast, VCAM-1 (another member of the immunoglobulin superfamily that interacts with β1 integrins of lymphocytes and monocytes) is absent in capillary endothelium, weakly expressed in EE and moderately expressed in endothelium of coronary arteries. ELAM-1 (belongs to the group of selectins and interacts with neutrophils) is also absent in capillary endothelium whereas EE, arteries and venules reveal a weak to moderate staining intensity. Endothelial cells of myocardial capillaries express constitutively MHCI and MHCII antigens whereas EE and endothelium of coronary arteries and venules only weakly express MHCI (see Table 2.2). Endothelium of coronary arteries do exhibit class II HLA-DR antigens.

Of the endothelial-specific markers, PECAM and EN4 showed a consistent and intense staining in all cardiac endothelial cells (Table 2.2).[72] Another endothelial marker, PAL-E was well ex-

Table 2.2. Distribution of adhesion molecules, MHC antigens and Von Willebrand factor in human cardiac endothelium and in large vessels (aorta and pulmonary artery)

	ICAM-1	VCAM-1	ELAM-1	MHC I	MHC II	DR	PECAM	PAL-E	vWF
EE	+	(+)	(+)	(+)	-	-	++	+	++
arteries	+	+	(+)	+	-	+	++	-	++
arterioles	+	(+)	-	+	-	(+).	++	-	++
capillaries	++	-	-	++	+	++	++	(+)	(+)
venules	+	(+)	+	+	-	(+)	++	+	++
large vessels	+	-	(+)	+	-	-	++	(+)	++

–, very weak and patchy staining or absence of staining; (+) very weak but uniform staining or patchy staining with moderate intensity; +, uniform staining with moderate intensity; ++, uniform and intense staining; ICAM-1, intercellular adhesion molecule-1; VCAM-1, vascular cell adhesion molecule-1 ; ELAM-1, endothelial-leukocyte adhesion molecule-1; MHC, major histocompatibility complex; DR, one of the members of the MHC II family; PAL-E, an endothelial marker; vWF, Von Willebrand factor. Data summarized from Page et al[72] and Taylor et al.[110]

pressed in capillaries and EE, but only very weakly in coronary arteries, whereas staining for von Willebrand factor (vWF) was intense in EE and in large vessels, it was comparatively weaker and less uniform in capillary endothelial cells.

Accordingly in the normal human heart, the various compartments of cardiac endothelium possess different phenotypic properties. These might either result from different microenvironments or might manifest adaptation to functional specializations. For example endothelium of myocardial capillaries may have evolved as effector cells for immune responses since it expresses molecules that are essential for interaction with lymphocytes. By contrast endothelium of larger vessels and of EE does not have these properties. Most of the above data were derived from human donor hearts where endothelium might have been activated as a consequence of the many medical procedures to prolong life of the donor.[72] A similar distribution of various adhesion molecules was, however, confirmed in hearts of rats (ICAM-1)[132] and rabbits (ICAM-1 and VCAM-1)[108] suggesting that the expression is constitutive and does not depend on immunologic upregulation. Pronounced upregulation of adhesion molecules, endothelial markers and/or MHC II antigens in cardiac endothelium has been observed both during rejection of transplanted hearts[43,108,110] and in dilated cardiomyopathy.[21,43]

3.3. Gap Junctions

Communication between EE and subjacent myocytes can be amplified further by intercellular communication between EE cells themselves, via gap junctions. EE cells are characterized by considerable overlap of the cellular borders[7,63] (Fig. 2.3A). Abundant well developed gap junctions are present in these overlapping zones, as was demonstrated by conventional *transmission electron microscopy* (TEM)[3,4,7] and by *en face confocal scanning laser microscopy* of immunolabeled connexin in EE[3] (Fig. 2.3). Connexins form a group of related proteins with different properties which appear to constitute the intercellular channels of gap junctions. At least four types of connexins have been identified of which Cx37, Cx40 and Cx43 were demonstrated in vascular endothelium. The considerable dissimilarity of the distribution of endothelial connexins in the vascular tree[17,74,76] as well as between different species is unclear. Whereas Cx43 staining neatly outlined the endothelial cell boundaries in ventricular EE and in aortic endothelium of the rat (Fig. 2.3C), staining lacked in endothelium of myocardial capillaries.[3] This latter observation confirms the absence of gap junctions in endothelium of capillaries in general.[50,95] Accordingly, whereas the absence of gap junctions in endothelium of capillaries suggests a more local regulatory function of these cells, the abundance of gap junctions in macrovascular endothelium and in EE is probably in accordance with an extensive sensor function. For example following activation of a single cell by a receptor mediated process, second messengers may pass the gap junctions, thereby interconnecting endothelial cells over large surface areas. The resulting

Fig.2.3 (right). (A-B) TEM micrographs of two serial sections showing the peripheral border of two overlapping EE cells in cat heart. Staining of the cell coat along the luminal cell surface and nearly the whole length of the intercellular cleft was obtained by tannic acid treatment after glutaraldehyde fixation.

(A) Large arrows indicate gap junctions. A protrusion of an interstitial cell penetrated the endothelial basal lamina but serial sections showed that close contacts were separated by narrow clefts (arrowheads). Scale bar: 250 nm.

(B) Small arrows delineate a gap junction that measures 0.6 μm in length. Note the nerve fiber (NF) in the endocardial interstitial tissue. Scale bar: 100 nm.

(C-D) En face confocal scanning laser microscopy (CSLM) of Cx43 immunostaining of EE in rat heart. Scale bar: 10 μm.

(C) Cx43-stained gap junctions (arrowheads) had variable sizes and outlined the cell borders of EE of the right ventricle. Nuclei of EE cells (Nu) were visible as dark spots.

(D) EE cells of the right atrio-ventricular valve possessed also many Cx43-stained gap junctions. Note the small size of the valvular EE cells compared to EE cells shown in C.

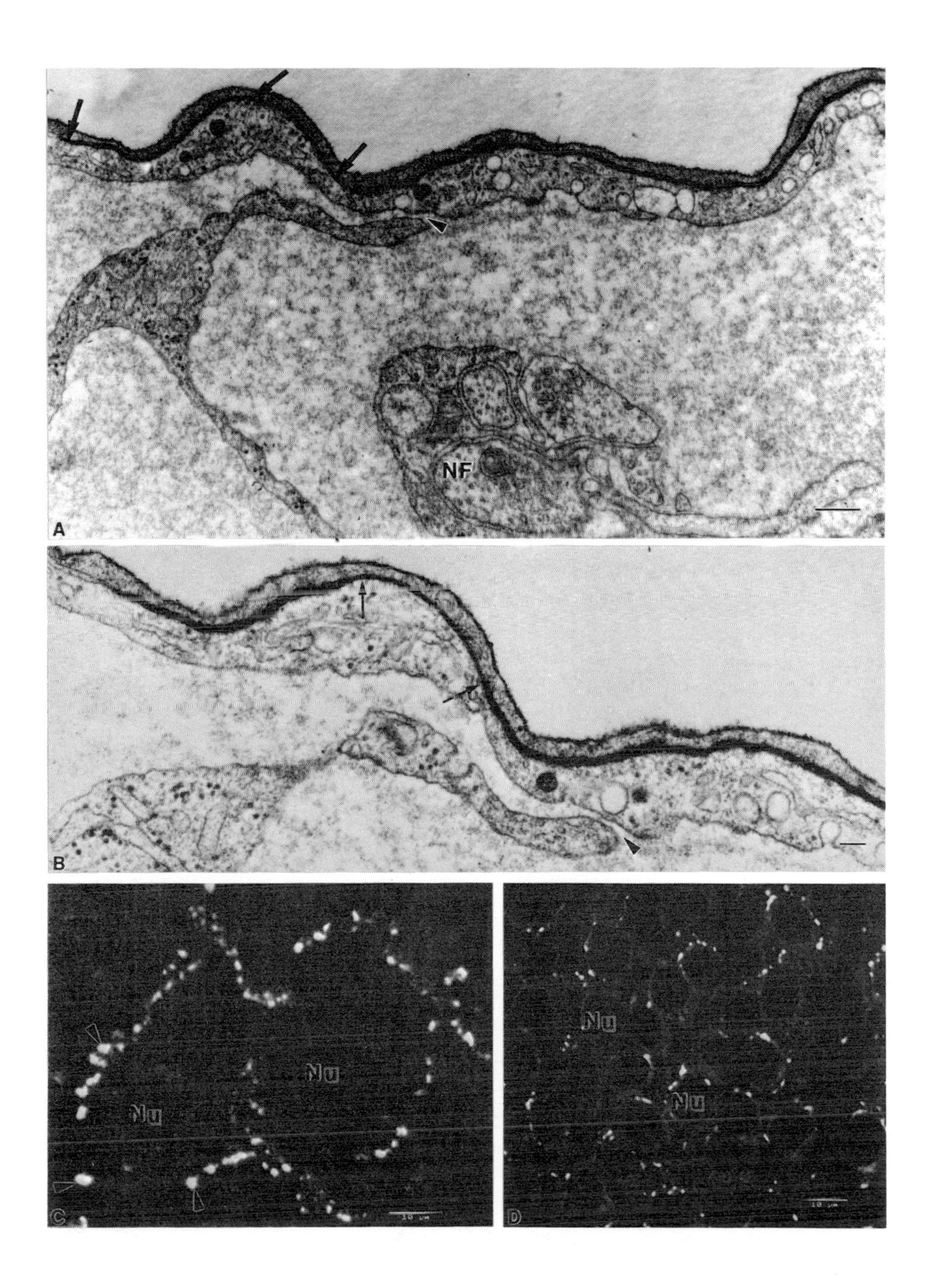
NF
A
B
Nu
Nu
C
Nu
Nu
D

"syncytial" behavior of the EE further amplifies its sensory function as well as the output capacity of endothelial factor release.

Could communication between EE cells and myocytes also be mediated through heterologous contacts via endocardial interstitial cells?[7] In vascular endothelium, myoendothelial junctions were reported to establish a direct communication between endothelial and smooth muscle cells.[20,78,102,109] Recently, electrical coupling was demonstrated between endothelium and smooth muscle cells of coronary vessels.[10] In addition, cardiac myocytes co-cultured with fibroblasts or epicardial mesothelial cells did form intercellular gap junctions and these heterologous gap junctions established electrical or dye coupling.[34,82] Yet in in vivo conditions, despite extensive morphological and immunohistochemical investigations no heterologous contacts between myocytes and fibroblasts were found.[31] In cat endocardium, not in rat, close contacts were noticed between EE and protrusions of fibroblast-like cells that penetrated the basal lamina of the EE (Fig. 2.3A). These contact zones did, however, not have membrane specializations resembling gap junctions.[3] Moreover, the space between the cell membranes of both cell types usually measured 10 nm, which is still 3 times larger than in genuine gap junctions. Heterologous gap junctional communication between EE and myocytes, via endocardial fibroblast-like cells, thus seems unlikely. The presence of small gap junctions in heterologous contacts cannot be excluded, however, by conventional TEM, as has been shown in smooth muscle of coronary arteries.[9]

In conclusion, the nonuniform distribution of various receptors, bioactive agents, adhesion molecules and gap junctions illustrates the heterogeneity of cardiac endothelium and affirms the sensory function of EE.

4. BLOOD COAGULATION, PLATELET AGGREGATION

Endothelial cells are involved in the regulation of coagulant, fibrinolytic and thrombotic mechanisms. Thrombomodulin is an intrinsic membrane receptor on endothelial cells that converts thrombin into an activator of protein C.[107] Activated protein C in combination with the coagulation factor protein S, which is also synthesized by endothelial cells, inactivates coagulation factors Va and VIIIa. The anticoagulant activity of thrombomodulin in EE might be impaired during eosinophilic endocarditis.[101] Eosinophilic endocarditis or Loeffler's endocarditis is characterized by prominent thromboembolism. Major basic protein and probably other substances released by eosinophils, accumulate on the EE surface and inhibit protein C generation, thereby promoting coagulation.[101] In addition, eosinophil peroxidase, another cationic granule protein, can bind to the EE surface and generate cytotoxic oxidants that damage the EE[100] and hence further increase thrombosis.

Platelet aggregation is inhibited by endothelium-derived NO and prostacyclin (PGI_2) which can act synergistically. EE produces NO[90,98] and high quantities of prostacyclin.[60] Blocking the synthesis of NO in cultured EE cells increased the adhesion of platelets.[97]

Endothelial cells also synthesize the von Willebrand factor (vWF or factor VIII-related antigen). This factor plays an important role in blood coagulation and platelet aggregation. As already shown in Table 2.2, capillary endothelium stains less intensively for vWF than endothelium of larger coronary vessels and EE in human hearts.[72] Only 30% of EN-4 positive vascular endothelial cells were positive for vWF. Endothelial cells secrete vWF via two pathways: a constitutive pathway, whereby low molecular molecules are secreted, and a regulated pathway, whereby large multimers (molecular weight up to 10-20 million dalton) are stored in granules and secreted after stimulation of the endothelial cell by thrombin, histamine and other secretagogues.[103,116] The large multimers bind most avidly to the platelets and to the extracellular matrix.[104] The storage granules of multimeric vWF appeared to be similar to the for a long known Weibel-Palade bodies.[118,122] Weibel-Palade bodies are rod-shaped granules which are 0.1 µm in diameter and up to 3 µm in length, and which contain a tubule-like material.[123] Besides multimeric vWF, Weibel-Palade bodies include P-selectin (PADGEM, GMP140 or CD62), which is an adhesion molecule for neutrophils and monocytes.[26,117]

Weibel-Palade bodies have a more pronounced heterogeneous distribution in the vascular system than vWF; in addition, they might be species dependent. In pigs, typical Weibel-Palade bodies were absent in endothelium of thoracic aorta, but present in myocardial capillaries and abundant in vena cava and pulmonary artery.[37] In other species, Weibel-Palade bodies were reported to be absent or rare in capillary endothelium.[50,119,123] In rat ventricular EE only few typical Weibel-Palade bodies were found and they were rather small, with a maximal diameter of 0.6 µm (Fig. 2.4B).[3] Many EE cells showed elliptical to spherical structures delineated by a membrane and filled with ring-like particles which have a diameter similar to tubules in typical Weibel-Palade bodies (Fig. 2.4B). It is at present not clear whether these organelles in rat EE are homologous to the atypical Weibel-Palade bodies present in all vascular endothelial cells of the pig.[37] *En face* confocal microscopy of EE stained for vWF, confirmed that rod-shaped granules, probably Weibel-Palade bodies, were not longer than 1 µm in ventricular EE of the rat (Fig. 2.4C).[3] Besides granules, a more amorphous juxta-nuclear staining was present in rat EE, since the used antibody also labeled the low molecular weight forms of vWF in endoplasmic reticulum and Golgi apparatus. In contrast, only minor juxta-nuclear labeling and many vWF positive granules, with a maximal length of 1.3 µm) were observed in endothelium of the pulmonary valve. In rat aortic endothelium, rod-shaped structures reached a length of 2.5 µm and much more amorphous staining was observed than in EE.[3] The differences in distribution of vWF and Weibel-Palade bodies are unclear. Probably, mechanical stress or other microenvironmental conditions can influence the structure of the Weibel-Palade bodies and the distribution of vWF.

Activation of endothelial cells can lead to an increased synthesis of vWF and is considered as one of the markers for dysfunctional endothelium.[11,58,84] Increased vWF staining of EE and other cardiac endothelial cells has been reported during rejection of transplanted hearts.[43,110] Activation of endothelial cells and increase of vWF staining is correlated with a hypertrophy of endoplasmic reticulum and Golgi apparatus,[77] the sites of vWF synthesis and multimerization. A considerable increase of endoplasmic reticulum and the Golgi complex has been noticed in EE injured by isoproterenol.[114] In normal EE cells, the Golgi apparatus has been described as small[79] or poorly developed.[63,114] In contrast, our results demonstrated the presence in EE of a well-developed Golgi apparatus consisting of up to six Golgi stacks and associated with many coated and uncoated vesicles (Fig. 2.4A, 2.7A).[3,15]

These observations suggest that EE cells are metabolically highly active cells.

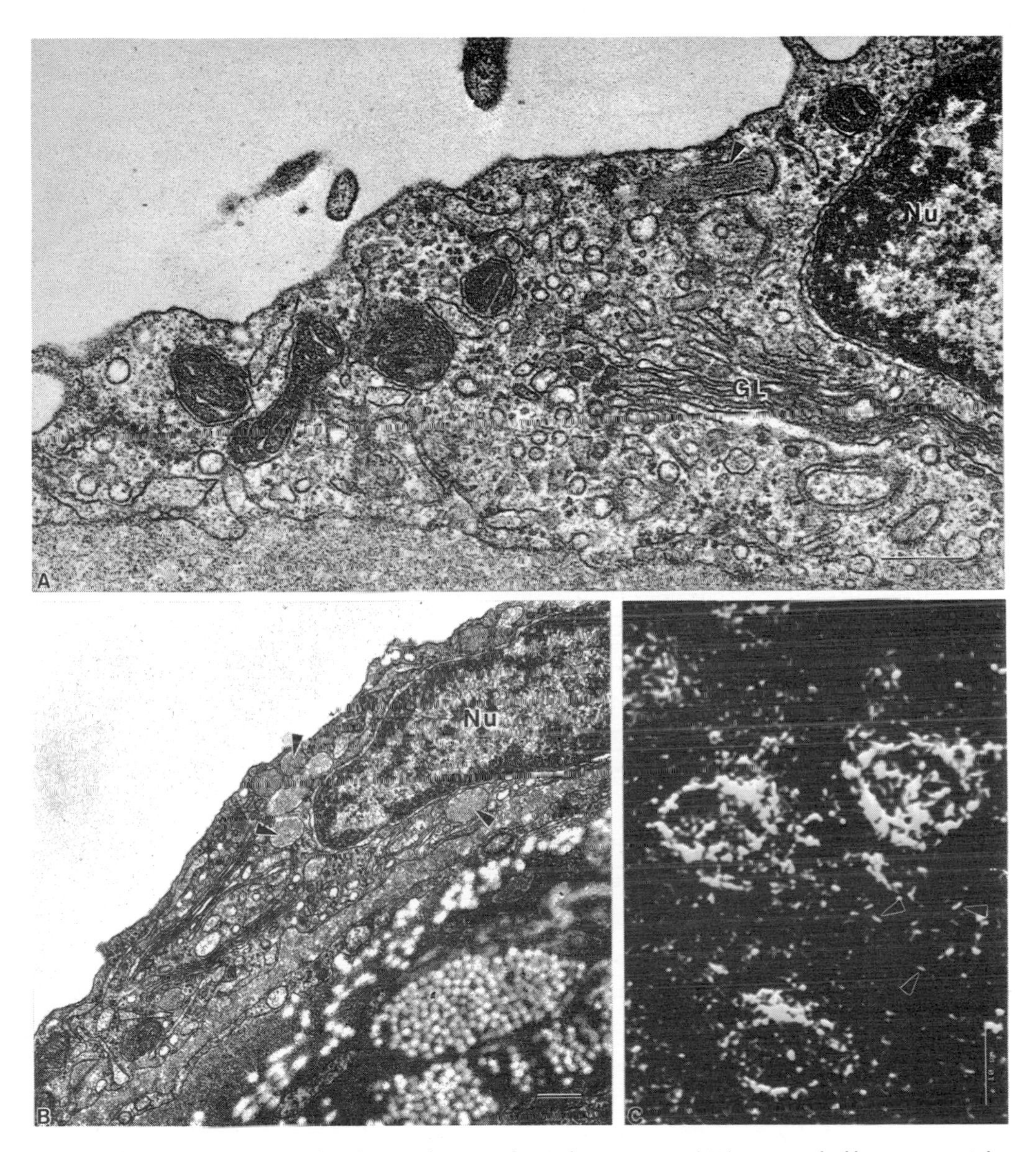

Fig. 2.4. (A-B) TEM micrographs of rat EE showing the Golgi apparatus (GL) surrounded by many vesicles close to the nucleus (Nu).

(A) A typical rod-shaped Weibel-Palade body (arrowhead) with tubular content was present near the luminal surface. Scale bar: 250 nm.

(B) Several granules (arrowheads) resembling Weibel-Palade bodies but filled with small ring-like structures were present in the vicinity of the nucleus. Scale bar: 100 nm.

(C) En face CSLM of ventricular EE in rat after immunostaining for the von Willebrand factor. Many small granules were present in the cytoplasm, some had a rod-like shape (arrowheads). The juxtanuclear staining represents the vWF in the vesicular system of the Golgi apparatus and endoplasmic reticulum. Scale bar: 10 µm.

5. PERMEABILITY AND CYTOSKELETON

5.1. Junctional Organization and Permeability

The endothelium forms a selective permeable barrier to the diffusion of molecules.[73,121] This transendothelial transport occurs mainly via two pathways: transcytosis and paracellular transport. In *transcytosis*, endothelial smooth vesicles can transport macromolecules by shuttling between the apical and basal cell membrane.[14,120] Alternatively, endothelial vesicles might fuse and form channels which allow the transport of macromolecules.[19,96] Uptake of macromolecules by smooth vesicles occurs by receptor-mediated mechanisms, as was demonstrated for albumin,[38,68] or by a fluid-phase mechanism for molecules which do not interact with the cell surface, such as native anionic ferritin.[75] Smooth vesicles can occupy up to 35% of the nonnuclear cell volume in endothelium of capillaries.[120] Considerable variation of vesicle density exists in capillaries of different organs, in the blood-brain barrier smooth vesicles being nearly absent.[25] In EE, the mean volume density of smooth vesicles was significantly lower than in endothelial cells of myocardial capillaries.[3] This finding is supported by Lupu and Simionescu[55] who found a density of 29-43 pits/μm^2 along the luminal side of valvular EE cells. This value is much lower than in capillaries of mouse diaphragm.[24] Whether transport by transcytosis is limited in EE is not known. Other features such as hydrostatic pressure can influence transcytotic transport.[29,81] Interestingly, in fishes with an exclusive spongy heart, lacking coronary vessels, the number of smooth vesicles appears to be much higher than in EE of higher mammals.

Paracellular transport occurs through the intercellular clefts and is limited by the presence of tight junctions or zonulae occludentes. Other structural features, such as the width of the intercellular clefts and the "en face" length of the clefts per μm^2, as well as the cleft depth from luminal to basal surface, are important determinants of paracellular permeability.[18] In addition, the glycocalyx in the intercellular clefts might function as a molecular sieve.[27,67] The complexity of tight junctions and intercellular clefts varies extensively between vascular endothelia of different organs[46] as well as between segments of the vascular tree within the same organ.[95]

In EE, tight junctions had a simple structure with one or two junctional contact points and occasionally open clefts without tight junctions (Fig. 2.5).[6,7] This organization of tight junctions appears to be similar as in endothelium of myocardial capillaries. In EE of cardiac valves, tight junctions were more complex with 3-4 tight junctional ridges on freeze-fracture images.[55] Regions of intercellular contacts in ventricular EE were usually complex with interdigatations and overlap of the peripheral borders[6,7,63] (Fig. 2.5A). The estimated depth of less complex and straightly sectioned intercellular clefts was longer than that of endothelial cells in myocardial capillaries.[6] The width of the intercellular cleft measured 20 nm, similar as in myocardial capillaries, and was constant even in long and complex clefts. The en face length of intercellular clefts per μm^2 was smaller in EE than in aortic endothelium and coronary microvascular endothelium.[6] Since permeability is proportional to length and width, and is inversely related to depth of intercellular clefts, paracellular transport through EE might be expected to be lower than through endothelium of myocardial capillaries.

However, permeability experiments demonstrated that intravenously injected LY-dextran 10000 (dextran with MW of 10000 and coupled to Lucifer Yellow) first penetrated the EE and only thereafter the myocardial capillary endothelium.[6] LY-dextran 40000 did not penetrate EE or endothelium in coronary vessels. In contrast, unbound Lucifer Yellow penetrated rapidly both EE and myocardial capillaries.[6] One reason for the increased permeation of LY-dextran 10000 through the EE could be the higher fluid pressure in the ventricular lumen than in the subendocardial myocardial capillaries, hence resulting in diffusion by convective transports. The fluorescent tracer penetrated the capillary endothelium in epicardial myocardium earlier than in subendocardial vessels. Permeation may, therefore, accord with differences in fluid pressure, being lower in subendocardial coronary vessels than in epicardial vessels.[81] The ultrastructural features of intercellular clefts in EE, such as the presence of deep and complex clefts and a the low value for length per μm^2, might be adaptations of the EE to limit transendothelial transport driven by the high hydrostatic ventricular pressures. A remarkable finding was the high permeability of large molecules such as LDL and horseradish peroxidase (HRP) in EE of aortic valves.[111] For as yet unknown reasons, the diffusion for LDL in heart valves was found to be 2 to 8 times faster than in aorta and veins.[111]

The above results with LY-dextran appear to be contradictory to a previous study where intravenously injected HRP was found to diffuse rapidly from the myocardial capillaries toward the ventricular lumen.[8] This unidirectional diffusion was explained by pressure gradients in the myocardial wall. HRP has a molecular

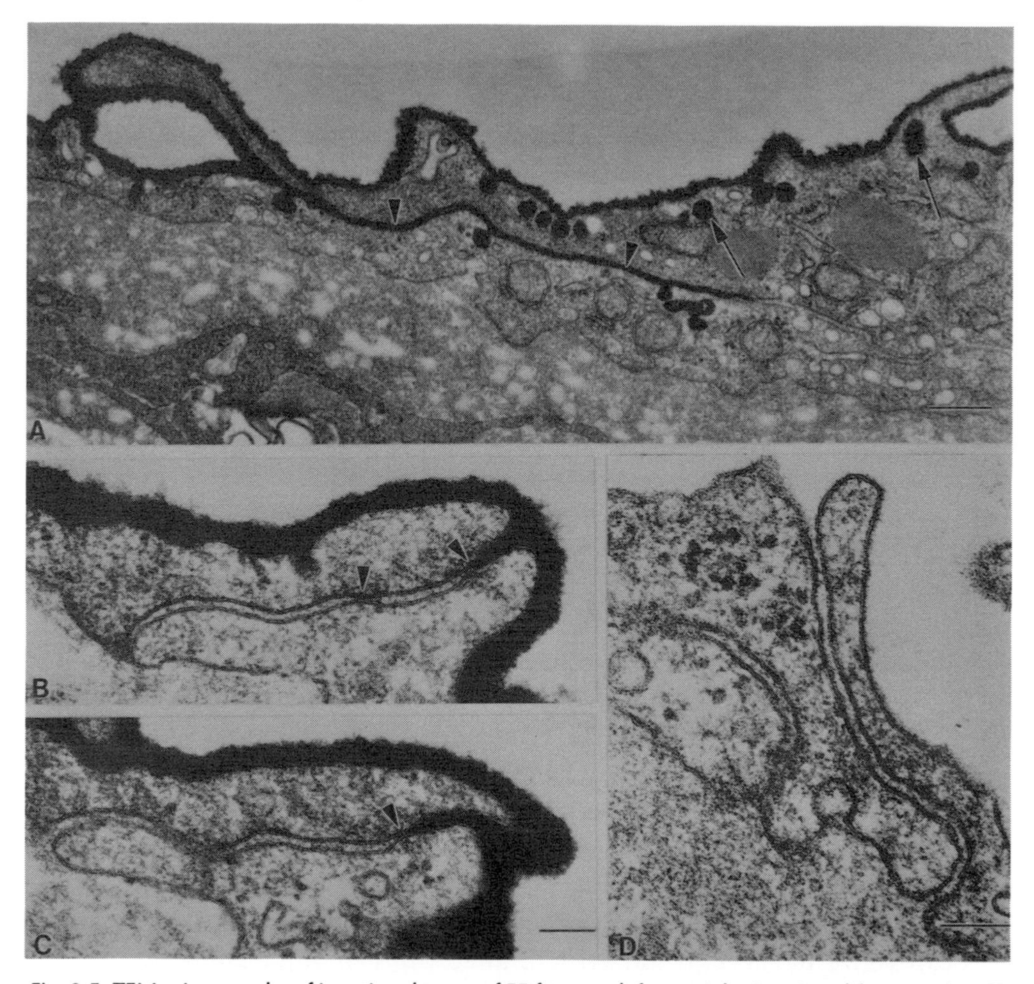

Fig. 2.5. TEM micrographs of junctional areas of EE from rat left ventricle. Reprinted from Andries LJ and Brutsaert DL, Cell Tissue Res 1994; 277:391-400, with permission from Springer-Verlag.

(A) Junctional area with extensive overlap between peripheral edges of 2 EE cells. The cell coat of the apical surface including that of apparent cytoplasmic vesicles (arrows), and part of the intercellular cleft (arrowheads) was densely stained by tannic acid. Scale bar: 250 nm.

(B-C) Micrographs of 2 sections from a series of serial sections showing an intercellular cleft between 2 EE cells. The cell coat of the luminal surface was well stained by tannic acid. Staining by tannic acid stopped at the first punctate junctional contact.

(B) Tight junction formed by 2 obliterating points (arrowheads) in the intercellular cleft.

(C) A single contact point (arrowhead) obliterated the intercellular cleft. Scale bar: 100 nm.

(D) Junctional area between 2 EE cells with open intercellular cleft. Scale bar: 100 nm.

weight comparable to LY-dextran 40000, but the commonly used Type II HRP consists of a mixture of at least two isoenzymes with a different charge.[94] Positively charged HRP can penetrate a monolayer of cells more easily than negatively charged HRP.[94,121] LY-dextrans have a net negative charge[6] and might thus possess different permeability properties from HRP. Accordingly, the net charge of macromolecules may be important in paracellular permeability. This also emphasizes the role of the negatively charged glycocalyx as proposed in the fiber matrix theory.[27]

In conclusion, the ultrastructure of the intercellular clefts in EE and the large surface area of EE cells may constitute an adaptation to limit diffusion driven by high hydrostatic pressure. The differences in transendothelial permeability, between EE and the subendocardial coronary endothelium, might assign characteristic electrochemical properties to the endocardium and subjacent myocardium.

5.2. Cytoskeleton and Cell Shape

Another important regulator of paracellular permeability is the peripheral actin band of the zonula adherens. In vitro experiments on endothelial cells have demonstrated that changes in paracellular permeability, evoked by thrombin and other vasoactive substances, were ATP and Ca^{++}-dependent and resulted from interactions between actin and myosin filaments.[83,88,129] Actin filaments are also organized as axially-aligned microfilament bundles or stress fibers. Stress fibers are numerous in endothelial cells of vessels subjected to high shear stress and turbulent flow, such as at flow dividers and branch sites in the aorta.[47,125] Stress fibers were absent in endothelium of veins, venules and capillaries.[32,44] Stress fibers play a role in maintaining the integrity of endothelium, presumably by enhancing cell-substratum adhesion (for a review see Gotlieb et al[40]) and by strengthening the endothelial cell surface.[126]

Similarly as in the endothelium of heart valves, arteries, and arterioles, earlier studies on EE cells in rat heart[32] and in bovine heart[128] suggested the presence of numerous stress fibers as a consequence of their exposure to high shear stress and flow turbulence in the heart. However, extensive *en face* confocal scanning laser microscopy of EE in rat heart demonstrated considerable regional differences in the distribution of stress fibers or microfilament bundles (MFB).[3,5]

In most EE cells of left and right ventricle in the rat, F-actin staining was confined to the cell periphery, thereby outlining a polygonal shape (Fig. 2.6A,C,D). In the apex, MFB's were found in the corner of trabecular ramifications and in EE cells covering very thin trabeculae. Elsewhere in the apex MFB's were rare. EE cells in the apex usually had a large surface area compared to, for example, aortic endothelial cells. Aortic endothelial cells have also a much more elongated cell shape than EE cells (Fig. 2.6F). In the right ventricle, MFB's in EE were always observed in a small area at the tendon end of the papillary muscles (Fig. 2.6B) and along the proximal edge of the atrioventricular valve. In these areas, EE cells had a small surface area and were frequently more elongated than EE in neighboring zones. Moreover, the MFBs in these cells were differently organized than in vascular endothelium.[3] Most of the MFBs were localized along the luminal surface of the cell, especially in EE of the atrioventricular valve, and they were frequently connected to the peripheral actin band. In aortic endothelial cells, typical stress fibers or MFBs are located along the basal surface of the cell, although some reports mentioned the presence of luminal stress fibers in vascular endothelium[125] and in valvular cells of chick embryos.[36] In the right ventricle, MFBs were also found in the outflow tract, but their occurrence was highly variable. In the left ventricle, fewer MFBs were present at the tendon end of papillary muscles than in right ventricle. The occurrence of MFBs was highly variable in the outflow tract of left ventricle. In some zones, few MFBs were present in EE (Fig. 2.6D) whereas in other regions many MFBs were observed (Fig. 2.6E). Consistently, a high number of MFBs with a disturbed pattern of peripheral actin bands was noticed in EE cells covering ridges that traversed the outflow tract. Most MFB's were oriented roughly parallel to blood flow.

As shear stress is expected to be high for EE cells in the outflow tract of left ventricle, it could explain the presence of MFB in this region. However, the presence of numerous MFBs in small EE cells along the tendon end and along the proximal border of the atrioventricular valve in right ventricle cannot be explained by shear stress alone. Both the tendon end and the proximal edge of the atrioventricular valve are highly elastic and have a different compliance than other parts of the cardiac wall. These elastic areas may undergo extensive changes in shape during the cardiac

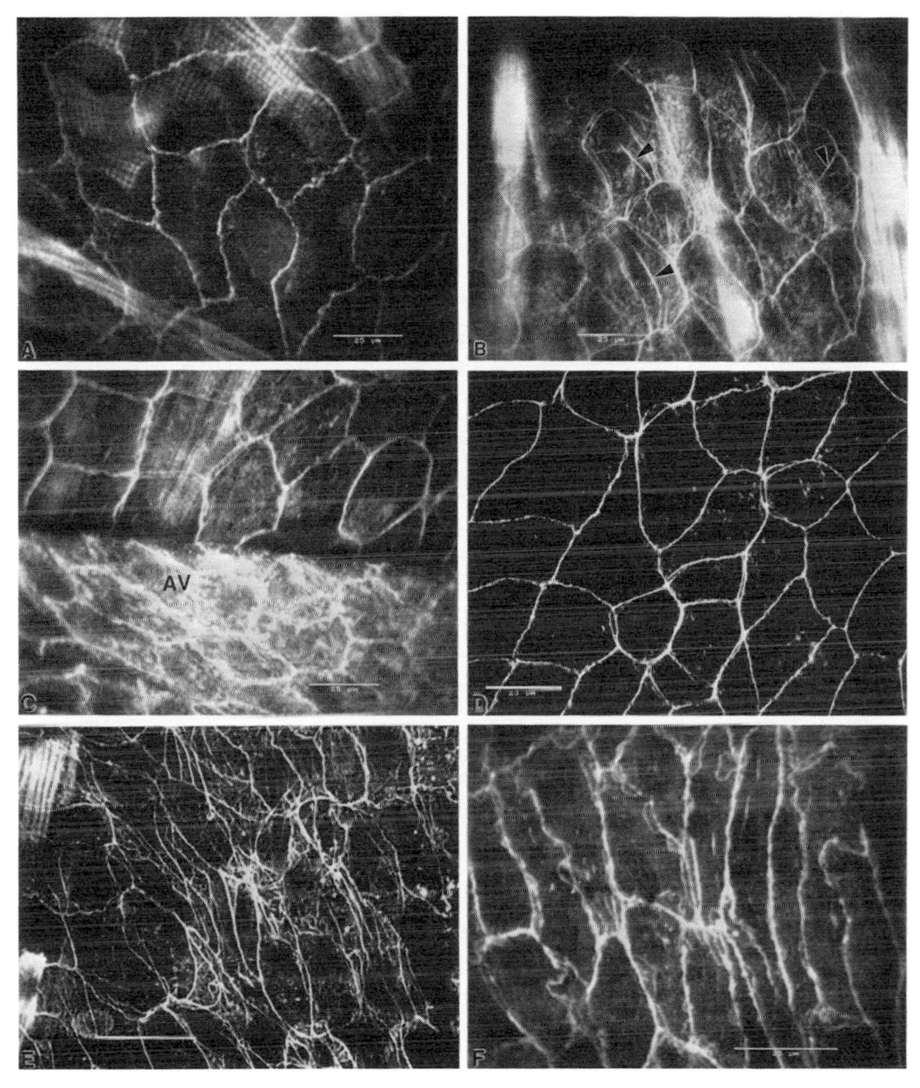

Fig 2.6. En face CSLM of F-actin stained by Bodipy-phallacidin of EE in rat right ventricle.

Scale bars: 25 μm.

(A) Near the septum end of papillary muscles, EE cells had distinct peripheral actin bundles.

(B) In addition to peripheral actin bundles, EE cells contained centrally located actin microfilament bundles (MFBs) (arrowheads) near the tendon end of papillary muscles.

(C) EE cells of the interventricular septum had a polygonal cell shape without a distinct orientation and without MFBs. These EE cells had a larger surface area than endothelial cells covering the atrioventricular valve (AV).

(D) EE cells in this area of the outflow tract of left ventricle had no MFBs. Reprinted from Andries LJ, Brutsaert DL, Cell Tissue Res 1993; 273:107-117 with permission from Springer-Verlag.

(E) In other areas of the outflow tract, the regular pattern of EE cells was disturbed by the presence of many MFBs.

(F) Endothelial cells in thoracic aorta were highly elongated and contained a variable number of MFBs.

cycle. Hence, EE cells in these areas might be subjected to more pronounced changes in cell shape than EE cells elsewhere in the ventricle. Previously, the arrangement of endothelial cells on aortic valves was also suggested to result from mechanisms other than shear stress.[30] Several in vitro experiments have demonstrated that vascular endothelial cells respond to cyclical stretch, similarly as in arterial vessels in vivo, by forming more stress fibers,[45] by a change in the orientation of stress fibers and by rearrangement of the pattern of endothelial cells.[28,93] Mechanical stress of EE cells on highly elastic substrata might thus influence the organization of F-actin and cell shape.

Besides F-actin, the cytoskeleton in most endothelial cells consists of microtubules and intermediate filaments. Microtubules are involved in vesicle transport and, in endothelial cells, intact microtubules are necessary for processing and secretion of von Willebrand factor.[99,103] Microtubules have also a structural function as has been demonstrated in cultured axons.[69] In aortic endothelium, centrioles, which are microtubule-organizing centers, are nonrandomly located in endothelial cells.[80] A similar nonrandom distribution was found in EE cells of aortic and pulmonary valves.[3,5] Microtubules were less aligned in ventricular EE than in aortic endothelium. In ventricular EE, centrioles were randomly distributed (Fig. 2.7B-D). The functional implications of these differences in the distribution of centrioles are at present not known.

Intermediate filaments in endothelial cells usually consist of vimentin which is ubiquitously present in mesenchymal cells.[106]

Fig. 2.7 (opposite). (A) TEM micrograph of the juxtanuclear region of a EE cell from rat left ventricle. Microtubules (arrowheads) radiated from a centriole (CN). Some small bundles of intermediate filaments (arrows) can be observed between components of endoplasmic reticulum and the well developed Golgi apparatus (GL). Nu: nucleus. Scale bar: 200 nm.

(B-E) Immunofluorescent staining of β-tubulin (B,C,D) or of the intermediate filament type vimentin (E) in endothelium. Reprinted from Andries LJ, Brutsaert DL, Cell Tissue Res 1994; 277:391-400 with permission from Springer-Verlag.

(B) Microtubules radiated from the centriolar regions (arrowheads) which were randomly oriented in EE of the right ventricle. Nu: nucleus. Scale bar: 10 μm.

(C) In many endothelial cells of the ventricular side of the pulmonary valve, centriolar regions (arrowheads) were located at the cardiac pole of each cell. Scale bar: 10 μm.

(D) Centriolar regions (arrowheads) in endothelial cells of the thoracic aorta were generally located at the cardiac pole of each cell. Microtubules appeared more closely packed in aortic endothelium than in EE. Scale bar: 25 μm.

(E) A network of fine vimentin filaments with some bundles (arrowheads) outlined the nuclei (Nu) in EE of the right ventricle. Scale bar: 10 μm.

The function of intermediate filaments is largely unknown.[106] Intermediate filaments probably contribute to the mechanical integrity of cells and tissues.[49] En face confocal microscopy of rat EE stained for vimentin showed a delicate network of filaments throughout the cytoplasm.[5] In many cells, bundles of vimentin filaments partly

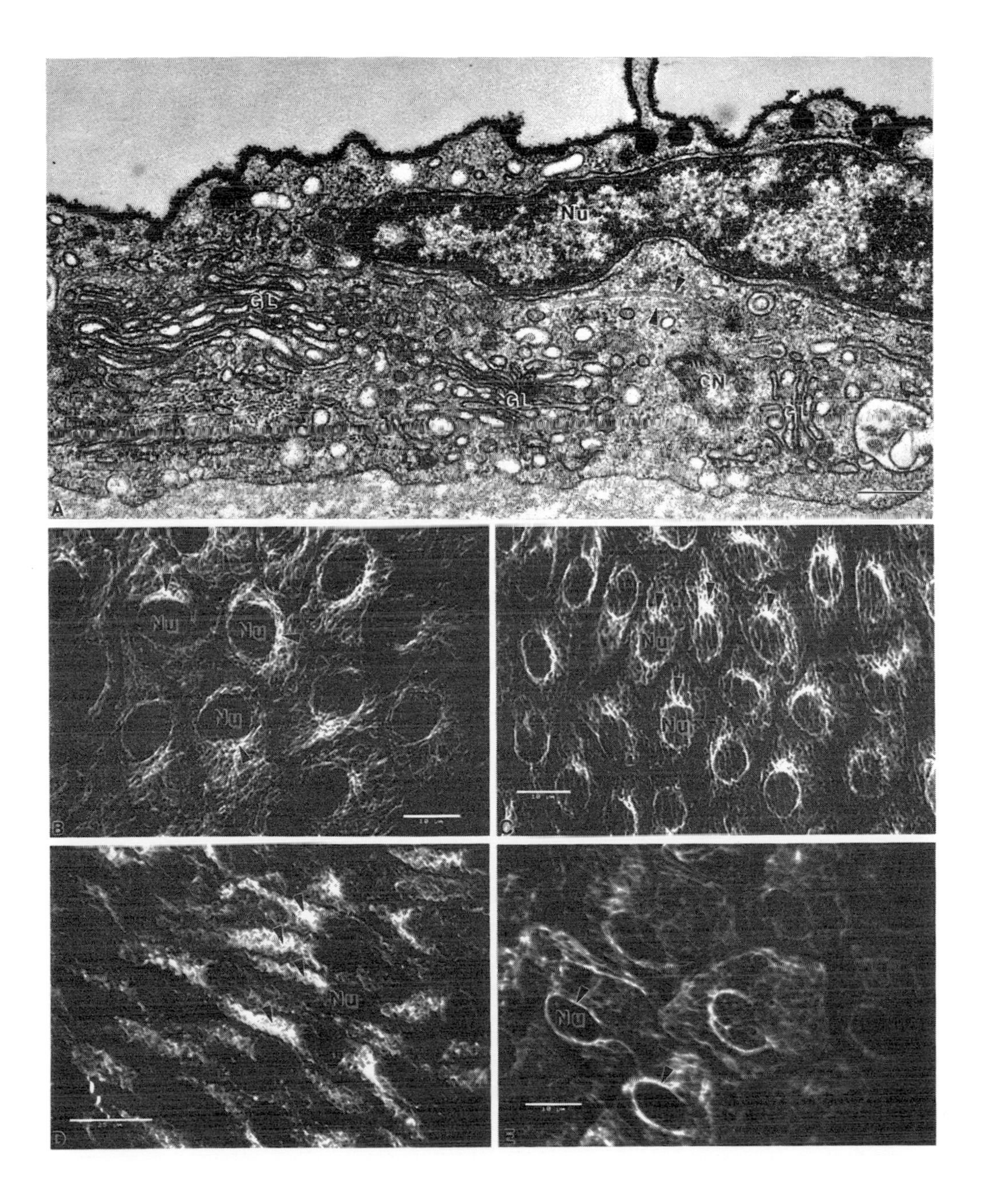

encircled the nucleus (Fig. 2.7D). In contrast to aortic endothelium of other species[12] perinuclear rings of vimentin filaments were never observed in EE cells or in aortic endothelium of rat. Shear stress as well as mechanical deformation of the ventricular wall during the cardiac cycle may affect the organization of actin filaments in EE. Differences in the cytoskeletal organization between EE and vascular endothelial cells may relate to differences in functional properties.

5.3. Actin Filaments as Indicators for Endothelial Injury

Alterations in hemodynamics and in the concentration of bioactive agents might lead to changes in the organization of the cytoskeleton in endothelial cells.[40] Reorganization of the cytoskeleton might alter endothelial functions and lead to endothelial dysfunction. Endothelial dysfunction may evolve in endothelial impairment and eventually lead to frank denudation of the vessel wall. In several models of chronic hypertension, an increase in the permeability of vascular endothelium was related to changes in the distribution of actin filaments.[35,39] In EE, high concentrations of isoproterenol induced alterations in the structure of junctional regions.[114] Chronically volume-loaded canine hearts observed with TEM showed elastofibrosis of the endocardium and an increase in MFBs and microvilli in EE.[57] These data suggest that changes in the actin filament system in endothelial cells might be an appropriate indicator of ongoing pathophysiological changes in these cells.

6. GENERAL CONCLUSION

The extensive trabeculated structure of the inner wall of the ventricles and the surface structure of the EE cells with numerous microvilli provide a high ratio of EE surface area to ventricular volume. The presence of several receptors and gap junctions in EE are in accordance with the presumed sensor function of the EE. The nonuniform distribution in cardiac endothelium of receptors, bioactive agents, adhesion molecules and gap junctions emphasizes the heterogeneity of the various endothelial cell types in the heart. This heterogeneity might either result from different microenvironments or might manifest adaptation to functional specializations and ontogenetical differences. EE cells are indeed formed very early in embryonic life and originate from the cardiogenic plate, whereas endothelial cells of the coronary vessels appear later and are derived from mesothelial cells in the epicardium. The deep and complex intercellular clefts and the large surface area of EE cells may constitute an adaptation to limit diffusion driven by high hydrostatic pressure in the heart. The differences in transendothelial permeability, between EE and the subendocardial coronary endothelium, might assign characteristic electrochemical properties to the endocardium and subjacent myocardium. Shear stress as well as mechanical deformation of the surface of the ventricular wall during the cardiac cycle may affect cell shape and the organization of the cytoskeleton in EE cells. Changes in mechanical stress during the pathogenesis of various cardiac diseases could induce a reorganization of the cytoskeleton, which could then alter endothelial functions eventually leading to endothelial dysfunction.

REFERENCES

1. Adyshirin-Zade EA, Gabain LI. Relief peculiarities of the internal surface of the cardiac ventricles and Viessen-Thebesian vessels. Arch Anat Histol Embryol 1984; 87:54-59.
2. Agnisola C, Tota B. Structure and function of the fish heart ventricle: flexibility and limitations. J Exp Zool 1995.
3. Andries LJ. Endocardial Endothelium: Functional Morphology. Austin: R.G. Landes Company, 1994:143.
4. Andries LJ, Brutsaert. Differences in functional structure between endocardial endothelium and vascular endothelium. J Cardiovasc Pharmacol 1991; 17(S3):S243-S246.
5. Andries LJ, Brutsaert DL. Endocardial endothelium in rat heart: cell shape and organization of the cytoskeleton. Cell Tissue Res 1993; 273:107-117.
6. Andries LJ, Brutsaert DL. Endocardial endothelium: junctional organization and permeability. Cell Tissue Res 1994; 277:391-400.
7. Anversa P, Giacomelli F, Wiener J. Intercellular junctions of rat endocardium. Anat Rec 1975; 183:477-484.
8. Anversa P, Giacomelli F, Wiener J et al. Permeability properties of ventricular endocardium. Lab Invest 1973; 28:728-734.
9. Bény JL, Connat JL. An electron microscopic study of smooth muscle cell dye coupling in the pig coronary arteries - Role of Gap Junctions. Circ Res 1992; 70:49-55.
10. Bény J-L, Pacicca C. Biderictional electrical communication between smooth muscle and endothelial cells in the pig coronary artery. Am J Physiol 1994; 266:H1465-H1472.
11. Blann A. Von Willebrand-factor and the endothelium in vascular disease. British Journal of Biomedical Science 1993; 50:125-134.
12. Blose SH, Meltzer DI. Visualization of the 10-nm filament vimentin

rings in vascular endothelial cells in situ. Exp Cell Res 1981; 135:299-309.
13. Boyd MT, Seward JB, Tajik AJ et al. Frequency and location of prominent left ventricular trabeculations at autopsy in 474 normal human hearts: implications for evaluation of mural thrombi by two-dimensional echocardiography. J Am Coll Cardiol 1987; 9:323-326.
14. Bruns RR, Palade GE. Studies on blood capillaries. II. Transport of ferritin molecules across the wall of muscle capillaries. J Cell Biol 1968; 37:277-299.
15. Brutsaert DL, Andries LJ. The endocardial endothelium Am J Physiol 1992; 263:H985-H1002.
16. Brutsaert DL, Meulemans AL, Sipido KR et al. Effects of damaging the endocardial surface on the mechanical performance of isolated cardiac muscle. Circ Res 1988; 62:357-366.
17. Bruzzone R, Haefliger JA, Gimlich RL et al. Connexin 40, a component of gap junctions in vascular endothelium, is restricted in its ability to interact with other connexins. Mol Biol Cell 1993; 4:7-20.
18. Bundgaard M, Frøkjaer-Jensen J. Functional aspects of the ultrastructure of terminal blood vessles: a quantitative study on consecutive segments of the frog mesenteric microvasculature. Microvasc Res 1982; 23:1-30.
19. Bundgaard M, Frøkjaer-Jensen J, Crone C. Endothelial plasmalemmal vesicles as elements in a system of branching invaginations from the cell surface. Proc Nat Acad Sci 1979; 76:6439-6442.
20. Burrig KF. The endothelium of advanced arteriosclerotic plaques in humans. Arterioscler Thromb 1991; 11:1678-89.
21. Caforio AL, Stewart JT, Bonifacio E et al. Inappropriate major histocompatibility complex expression on cardiac tissue in dilated cardiomyopathy. Relevance for autoimmunity? J Autoimmun. 1990; 3:187-200.
22. Candiollo L. The fine structure of the endocardial endothelium. Zeitschrift für Zellforschung 1963; 61:486-492.
23. Casley-Smith JR. Comparative fine structure of the microvasculature and endothelium. Adv Microcirc 1980; 9:1-44.
24. Casley-Smith JR. Vesicular form and fusion as revealed by freeze-immobilization and stereoscopy of semi-thin sections. Prog appl Microcirc 1985; 9:6-20.
25. Cervos-Navarro J, Kannuki S, Nakagawa Y. Blood-brain barrier (BBB). Review from morphological aspect. Histol Histopath 1988; 3:203-213.
26. Collins PW, Macey MG, Cahill MR et al. Von Willebrand factor release and p-selectin expression is stimulated by thrombin and trypsin but not IL-1 in cultured human endothelial cells. Thromb Haem 1993; 70:346-350.
27. Curry FE, Michel CC. A fiber matrix model of capillary permeability. Microvasc Res 1980; 20:96-99.

28. Dartsch PC, Betz E. Response of cultured endothelial cells to mechanical stimulation. Basic Res Cardiol 1989; 84:268-281.
29. DeFouw DO. Morphometric studies of endothelial vesicles of alveolar vessels in edematous lungs. Prog appl Microcirc 1985; 9:67-79.
30. Deck D. Endothelial cell orientation on aortic valve leaflets. Cardiovasc Res 1986; 20:760-767.
31. De Mazière AMGL, van Ginneken ACG, Wilders R et al. Spatial and functional relationship between myocytes and fibroblasts in the rabbit sionatrial node. J Mol Cell Cardiol 1992; 24:567-578.
32. Drenckhahn D. Cell motility and cytoplasmic filaments in vascular endothelium. Prog appl Microcirc 1983; 1:53-70.
33. Edanaga M. A scanning electron microscope study on the endothelium of vessels. II. Fine surface structure of the endocardium in normal rabbits and rats. Arch histol jap 1975; 37:301-312.
34. Eid H, Larson DM, Springhorn JP et al. Role of epicardial mesothelial cells in the modification of phenotype and function of adult rat ventricular myocytes in primary coculture. Circ Res 1992; 71:40-50.
35. Gabbiani G, Elemer G, Guelpa C et al. Morphological and functional changes of the aortic intima during experimental hypertension. Am J Pathol 1979, 96:399-414.
36. Garcia-Martinez V, Hurle JM. Cell shape and cytoskeletal organization of the endothelial cells of the semilunar heart valves in the developing chick. Anat Embryol 1986; 174:83-89.
37. Gebrane-Younes J, Drouet L, Caen JP et al. Heterogeneous distribution of Weibel-Palade bodies and von Willebrand factor along the porcine vascular tree. Am J Pathol 1991; 139:1471-84.
38. Ghitescu L, Fixman A, Simionescu M et al. Specific binding sites for albumin restricted to plasmalemmal vesicles of continuous capillary endothelium: receptor-mediated transcytosis. J Cell Biol 1986; 102:1304-1311.
39. Giacomelli F, Wiener J, Spiro D. Cross-striated arrays of filaments in endothelium. J Cell Biol 1970; 45:188-192.
40. Gotlieb AI, Langille L, Wong MKK et al. Biology of disease. Structure and function of the endothelial cytoskeleton. Lab Invest 1991; 65:123-137.
41. Grant RT, Regnier M. The comparative anatomy of the cardiac coronary vessels. Heart 1926; 13:285-317.
42. Hemsen A, Lundberg JM. Presence of endothelin-1 and endothelin-3 in peripheral tissues and central nervous system of the pig. Regul Pept 1991; 36:71-83.
43. Hengstenberg C, Rose ML, Olsen EG et al. Immune response to the endothelium in myocarditis, dilated cardiomyopathy and rejection after heart transplantation. Eur Heart J 1991; 12 Suppl D:144-146.
44. Herman IM, Pollard TD, Wong AJ. Contractile proteins in endot-

helial cells. Annals New York Academy of Sciences 1982; 401:50-60.
45. Iba T, Sumpio BE. Morphological response of human endothelial cells subjected to cyclic strain in vitro. Microvasc Res 1991; 42:245-254.
46. Jain RK. Transport of molecules across tumor vasculature. Cancer Metastasis Rev 1987; 6:559-593.
47. Kim DW, Langille BL, Wong MKK et al. Patterns of endothelial microfilament distribution in the rabbit aorta in situ. Circ Res 1989; 64:21-31.
48. Klein W, Böck P. Elastica-positive material in the atrial endocardium. Light and electron microscopic identification. Acta Anat 1983; 116:106-113.
49. Klymkowsky MW, Bachant JB, Domingo A. Functions of intermediate filaments. Cell Motility and Cytoskeleton 1989; 14: 309-331.
50. Kumar S, West DC, Ager A. Heterogeneity in endothelial cells from large vessels and microvessels. Differentiation 1987; 36:57-70.
51. Kuprijanov VV. [Vascular endothelium (review). I. General morphology. 2B: phylogenesis of the vascular endothelium. Gegenbaurs Morphol Jahrb 1990; 136:201-217.
52. Lannigan RA, Zaki SA. Ultrastructure of the normal atrial endocardium. Brit Heart J 1966; 28:785-795.
53. Lemanski LF, Fitts EP, Marx BS. Fine structure of the heart in the Japanese medaka, Oryzias latipes. J Ultrastruct Res 1975; 53:37-65.
54. Lombès M, Oblin ME, Gasc JM et al. Immunohistochemical and biochemical evidence for a cardiovascular mineralocorticoid receptor. Circ Res 1992; 71:503-510.
55. Lupu F, Simionescu M. Organization of the intercellular junctions in the endothelium of cardiac valves. J Submicrosc Cytol 1985; 17:119-132.
56. Marron K, Wharton J, Sheppard Mn et al. Human endocardial innervation and its relationship to the endothelium: an immunohistochemical, and quantitative study. Cardiovasc Res 1994; 28:1490-1499.
57. Masuda H, Kawamura K, Tohda K et al. Endocardium of the left ventricle in volume-loaded canine heart. Acta Pathol Jpn 1989; 39:111-120.
58. Matucci-Cerinic M, Jaffa A, Kahaleh B. Angiotensin: an in vivo and in vitro marker of endothelial injury. J Lab Clin Med 1992; 120:428-433.
59. McMillan JB, Lev M. The aging heart. I. Endocardium. J Gerontol 1959; 14:268-283.
60. Mebazaa A, Martin LD, Robotham JL et al. Right and left ventricular cultured endocardial endothelium produces prostacyclin and PGE2. J Mol Cell Cardiol 1993; 25:245-248.
61. Mebazaa A, Mayoux E, Maeda K et al. Paracrine effects of endocardial endothelial cells on myocyte contraction mediated via

endothelin. Am J Physiol 1993; 265:H1841-1846.
62. Mebazaa A, Wetzel R, Cherian M et al. Comparison between endocardial and great vessel endothelial cells: morphology, growth, and prostaglandin release. Am J Physiol 1995; 268:H250-H259.
63. Melax H, Leeson TS. Fine structure of the endocardium in adult rats. Cardiovasc Res 1967; 1:349-355.
64. Meulemans AL, Andries LJ, Brutsaert DL. Endocardial endothelium mediates positive inotropic response to alpha1-adrenoreceptor agonist in mammalian heart. J Mol Cell Cardiol 1990; 22:667-685.
65. Meulemans AL, Andries LJ, Brutsaert DL. Does endocardial endothelium mediate positive inotropic response to angiotensin I and angiotensin II? Circ Res 1990; 66:1591-1601.
66. Meulemans AL, Sipido KR, Sys SU et al. Atriopeptin III induces early relaxation of isolated mammalian papillary muscle. Circ Res 1988; 62:1171-1174.
67. Michel CC. Capillary exchange. In: Seldin DW, Giebish G, ed. The kidney: physiology and pathophysiology. New York: Raven Press, 1992:61-91.
68. Milici JA, Watrous NE, Stukenbrok H et al. Transcytosis of albumin in capillary endothelium. J Cell Biol 1987; 105:2603-2612.
69. Miller RH, Lasek RJ, Katz M. Preferred microtubules for vesicle transport in lobster axons. Science 1987; 235:220-222.
70. Molenaar P, O'Reilly G, Sharkey A et al. Characterization and localization of endothelin receptor subtypes in the human atrioventricular conducting system and myocardium. Circ Res 1993; 72:526-538.
71. Ostadal B. Phylogenetic and ontogenetic development of the terminal vascular bed in the heart muscle and its effect on the development of experimental cardiac necrosis. II Annual Meeting of the International Study Group for Research in Cardiac Metabolism 1969:111-132.
72. Page C, Rose M, Yacoub M et al. Antigenic heterogeneity of vascular endothelium. Am J Pathol 1992; 141:673-683.
73. Pappenheimer JR, Renkin JR, Borrero LM. Filtration, diffusion and molecular sieving through peripheral capillary membranes. A contribution to the pore theory of capillary permeability. Am J Physiol 1951; 167:13-46.
74. Pepper MS, Montesano R, Elaoumari A et al. Coupling and connexin-43 expression in microvascular and large vessel endothelial cells. Am J Physiol 1992; 262:C1246-C1257.
75. Pietra GG, Johns L, Byrnes W et al. Inhibition of adsorptive endocytosis and transcytosis in pulmonary microvessels. Lab Invest 1988; 59:683-691.
76. Reed KE, Westphale EM, Larson DM et al. Molecular cloning and functional expression of human connexin 37, an endothelial cell gap junction protein. J Clin Invest 1993; 91:997-1004.
77. Reidy MA, Chopek M, Chao S et al. Injury induces increase of

Von Willebrand factor in rat endothelial cells. Am J Pathol 1989; 134:857-864.
78. Rhodin JAG. The ultrastructure of mammalian arterioles and precapillary sphincters. J Ultrastruct Res 1967; 18:181-223.
79. Rhodin JAG. Histology. A text and Atlas. London: New-York Oxford University Press, 1974:803.
80. Rogers KA, Kalnins V. Comparison of the cytoskeleton in aortic endothelial cells in situ and in vitro. Lab Invest 1983; 49:650-654.
81. Rona G, Hüttner I, Boutet M. Microcirculatory changes in myocardium with particular reference to catecholamine-induced cardiac muscle cell injury. In: Mussen H, ed. Handbuch der allgemeinen Pathologie III/7 Microzirkulation/Microcirculation . Berlin/Heidelberg/New York: Springer Verlag, 1977:791-888.
82. Rook MB, van Ginneken AC, de Jonge B et al. Differences in gap junction channels between cardiac myocytes, fibroblasts, and heterologous pairs. Am J Physiol 1992; 263:C959-77.
83. Rotrosen D, Gallin JI. Histamine type I receptor occupancy increases endothelial cytosolic calcium, reduces F-actin, and promotes albumin diffusion across cultured endothelial monolayers. J Cell Biol 1986; 103:2379-2387.
84. Rubanyi GM. The role of endothelium in cardiovascular homeostasis and diseases. J Cardiovasc Pharmacol 1993; 22:S1-14.
85. Ryan U, Slavkin H, Revel J-P et al. Conference report: cell-to-cell interactions in the developing lung. Tissue & Cell 1984; 16:829-841.
86. Sarphie TG, Allen DJ. Scanning and transmission electron microscopy of normal and methotrexate-treated endocardial cell populations in dogs. J Submicr Cytol 1978; 10:15-25.
87. Scheuermann DW. Ultrastructure of ventricular cardiac muscle of Rana temporaria. Advances in Anatomy, Embryology and Cell Biology 1974; 48:7-69.
88. Schnittler HJ, Wilke A, Gress T et al. Role of actin and myosin in the control of paracellular permeability in pig, rat and human vascular endothelium. J Physiol 1990; 431:379-401.
89. Schoemaker IE, Meulemans AL, Andries LJ et al. Role of the endocardial endothelium in the positive inotropic action of vasopressin. Am J Physiol 1990; 259:H1148-H1151.
90. Schulz R, Smith JA, Lewis MJ et al. Nitric oxide synthase in cultured endocardial cells of the pig. Br J Pharmacol 1991; 104:21-24.
91. Shah AM, Andries LJ, Meulemans AL et al. Endocardium modulates inotropic response to 5-hydroxytryptamine. Am J Physiol 1989; 257:H1790-H1797.
92. Shah AM, Smith JA, Lewis MJ. The role of endocardium in the modulation of contraction of isolated papillary muscles of the ferret. J Cardiovasc Pharmacol 1991; 17 (S3):S251-S257.
93. Shirinsky VP, Antonov AS, Birukov KG et al. Mechano-chemical control of human endothelium orientation and size. J Cell Biol

1989; 109:331-339.
94. Sibley CP, Bauman KF, Firth JA. Molecular charge as a determinant of macromolecular permeability across the fetal capillary endothelium of the guinea-pig placenta. Cell Tissue Res 1983; 229:365-377.
95. Simionescu M, Simionescu N, Palade GE. Segmental differentiation of cell junctions in the vascular endothelium. J Cell Biol 1975; 67:863-885.
96. Simionescu N, Simionescu M, Palade GE. Permeability of muscle capillaries to small heme peptides. Evidence for the existence of patent transendothelial channels. J Cell Biol 1975; 64:586-607.
97. Siney L, Lewis MJ. Endothelium-derived relaxing factor inhibits platelet adhesion to cultured porcine endocardial endothelium. Eur J Pharmacol 1992; 229:223-226.
98. Siney L, Lewis MJ. Nitric oxide release from porcine mitral valves. Cardiovasc Res 1993; 27:1657-1661.
99. Sinha S, Wagner DD. Intact microtubules are necessary for complete processing, storage and regulated secretion of von Willebrand factor by endothelial cells. Eur J Cell Biol 1987; 43:377-83.
100. Slungaard A, Mahoney JR. Bromide-dependent toxicity of eosinophil peroxidase for endothelium and isolated working rat hearts: a model for eosinophilic endocarditis. J Exp Med 1991; 173:117-26.
101. Slungaard A, Vercellotti GM, Tran T et al. Eosinophil cationic granule proteins impair thrombomodulin function. A potential mechanism for thromboembolism in hypereosinophilic heart disease. J Clin Invest 1993; 91:1721-1730.
102. Sosa-Melgarejo JA, Berry CL, Dodd S. Myoendothelial contacts in the small arterioles of human kidney. Virchows Archiv A Pathol Anat 1988; 413:183-187.
103. Sporn LA, Marder VJ, Wagner DD. Differing polarity of the constitutive and regulated secretory pathways for Willebrand factor in endothelial cells. J Cell Biol 1989; 108:1283-1289.
104. Sporn LA, Marder VJ, Wagner DD. Von Willebrand factor released from Weibel-Palade bodies binds more avidly to extracellular matrix than that secreted constitutively. Blood 1987; 69: 1531-1534.
105. Staley NA, Benson ES. The ultrastructure of frog ventricular cardiac muscle and its relationship to mechanisms of excitation-contraction coupling. J Cell Biology 1968; 38:99-114.
106. Steinert P, Jones JCR, Goldman RD. Intermediate filaments. J Cell Biol 1984; 99:22s-27s.
107. Stern D, Brett J, Harris K et al. Participation of endothelial cells in the protein C-protein S anticoagulant pathway: the synthesis and release of protein S. J Cell Biol 1986; 102:1971-1978.
108. Tanaka H, Sukhova GK, Swanson SJ et al. Endothelial and smooth muscle cells express leukocyte adhesion molecules heterogeneously during acute rejection of rabbit cardiac allografts. Am J Pathol 1994;

144:938-951.
109. Taugner R, Kirchheim H, Forssmann WG. Myoendothelial contacts in glomerular arterioles and in renal interlobular arteries of rat, mouse and Tupaia belangeri. Cell Tissue Res 1984; 235: 319-325.
110. Taylor PM, Rose ML, Yacoub MH et al. Induction of vascular adhesion molecules during rejection of human cardiac allografts. Transplantation 1992; 54(3):451-457.
111. Tompkins RG, Schnitzer JJ, Yarmush ML. Macromolecular transport within heart valves. Circ Res 1989; 64:1213-1223.
112. Tota B. Vascular and metabolic zonation in the ventricular myocardium of mammals and fishes. Comp Biochem Physiol 1983; 76A:423-437.
113. Tota B, Cimini V, Salvatore G et al. Comparative study of the arterial and lacunary systems of the ventricular myocardium of elasmobranch and teleost fishes. Am J Anat 1983; 167:15-32.
114. Turcotte H, Bazin M, Boutet M. Junctional complexes in regenerating endocardium. J Ultrastruct Res 1982; 79:133-141.
115. Ursell PC, Mayes M. The majority of nitric oxide synthase in pig heart is vascular and not neural. Cardiovasc Res 1993; 27: 1920-1924.
116. Van Buul-Wortelboer MF, Brinkman H-JM, Reinders JH et al. Polar secretion of von Willebrand factor by endothelial cells. Biochim Biophys Acta 1989; 1011:129-139.
117. Wagner DD. The Weibel-Palade body - the storage granule for von willebrand factor and p-selectin. Thrombosis and Haemostasis 1993; 70:105-110.
118. Wagner DD, Olmsted JB, Marder VJ. Immunolocalization of Von Willebrand protein in Weibel-Palade bodies of human endothelial cells. J Cell Biol 1982; 95:355-360.
119. Wagner RC. Endothelial cell embryology and growth. Adv Microcirc 1980; 9:45-75.
120. Wagner RC, Casley-Smith JR. Endothelial vesicles. Microvasc Res 1981; 21:267-298.
121. Ward BJ, Bauman KF, Firth JA. Interendothelial junctions of cardiac capillaries in rats: their structure and permeability properties. Cell Tissue Res 1988; 252:57-66.
122. Warhol MJ, Sweet JM. The ultrastructural localization of von Willebrand factor in endothelial cells. Am J Pathol 1984; 117:310-315.
123. Weibel ER, Palade GE. New cytoplasmic components in arterial endothelia. J Cell Biol 1964; 23:101-112.
124. Wharton J, Rutherford RA, Gordon L et al. Localization of endothelin binding sites and endothelin-like immunoreactivity in human fetal heart. J Cardiovasc Pharmacol 1991; 17 S 7:S378-384.
125. White GE, Fujiwara K. Expression and intracellular distribution of stress fibers in aortic endothelium. J Cell Biol 1986; 103:63-70.

126. White GE, Gimbrone MA, Fujiwara K. Factors influencing the expression of stress fibers in vascular endothelial cells in situ. J Cell Biol 1983; 97:417-424.
127. Wilcox JN, Augustine A, Goeddel DV et al. Differential regional expression of three natriuretic peptide receptor genes within primate tissues. Mol Cell Biol 1991; 11:3454-3462.
128. Wong AJ, Pollard TD, Herman IM. Actin filament stress fibers in vascular endothelial cells in vivo. Science 1983; 219:867-869.
129. Wysolmerski RB, Lagunoff D. Inhibition of endothelial cell retraction by ATP depletion. Am J Pathol 1988; 132:28-37.
130. Yamada H, Fabris B, Allen AM et al. Localization of angiotensin converting enzyme in rat heart. Circ Res 1991; 68:141-149.
131. Yamauchi A. Fine structure of the fish heart. In: Hearts and heart-like organs. Vol. 1: Academic Press, 1980:119-143.
132. Yamazaki T, Seko Y, Tamtani T et al. Expression of intercellular adhesion molecule-1 in rat heart with ischemia/reperfusion and limitation of infarct size by treatment with antibodies against cell adhesion molecules. Am J Pathol 1993; 143:410-418.

CHAPTER 3

Endothelial Control of Myocardial Performance

Gilles De Keulenaer and Stanislas Sys

Ventricular performance of the heart relies largely on autoregulatory mechanisms. Autoregulation is accomplished through intrinsic feedback by the loading conditions and the length of the muscle fibers (heterometric autoregulation), and through extrinsic feedback by the activity of the neurohumoral system (homeometric autoregulation). In 1988, Brutsaert et al reported that selective damage of the endocardial endothelium (EE) in isolated cat and rat cardiac muscle resulted in a typical modulation of the twitch contraction[3] and hypothesized that endothelial cells from the endocardium might be involved in the regulation of ventricular function. This observation was subsequently confirmed by other investigators in various animal species[12,13,26] and in the intact heart,[8,9] and has been extended to the endothelium of the coronary (micro)vasculature.[13] Moreover, cardiac endothelial cells have now been shown to release several inotropic substances, such as endothelin,[15] prostaglandins[16] and nitric oxide.[24] Although it is not yet fully established whether other substances or mechanisms participate and how they are regulated, there is now compelling evidence to complete, or at least, to change our traditional picture of cardiac performance (Fig. 3.1).[5] In addition to the above autoregulatory mechanisms, cardiac endothelial cells from the endocardium and the microvasculature directly control the underlying

Endocardial Endothelium: Control of Cardiac Performance, edited by Stanislas U. Sys and Dirk L. Brutsaert.

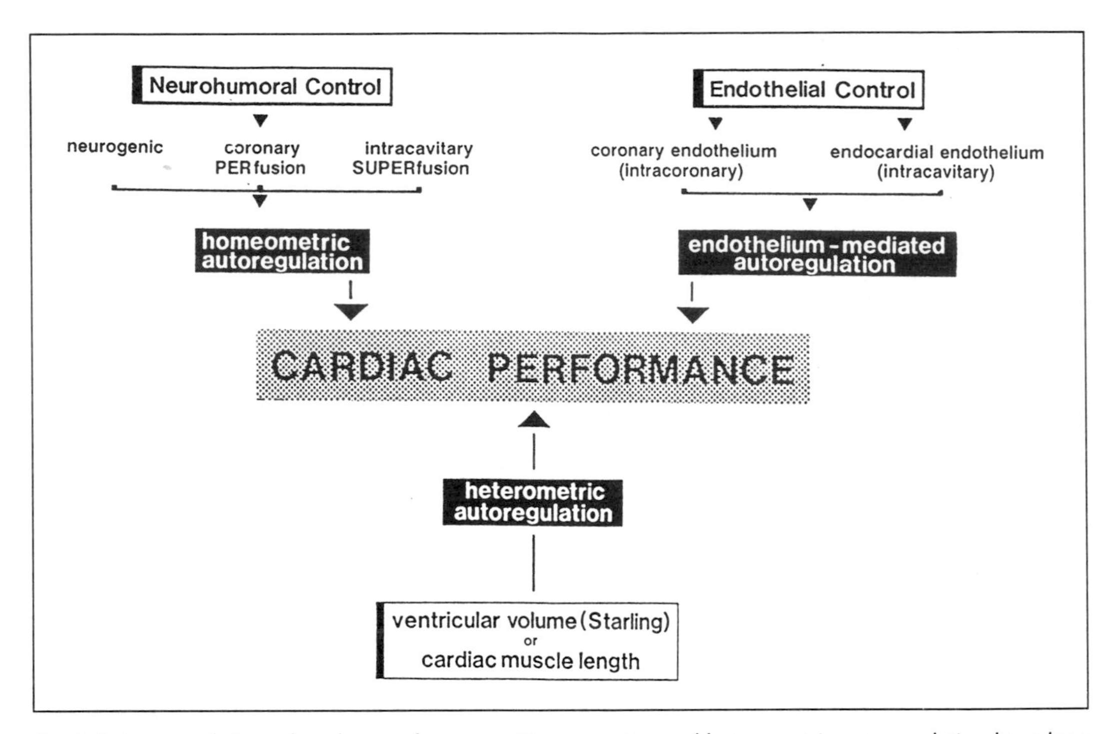

Fig. 3.1. Autoregulation of cardiac performance. Homeometric and heterometric autoregulation have long been assumed to be the sole intrinsic and extrinsic controlling mechanisms of cardiac performance. Autoregulation by endocardial and by (micro)vascular endothelial cells adds a unique and independent type of controlling mechanism to the system. Development of endocardial endothelium (EE) precedes, phylogenetically and embryologically, differentiation and contraction of cardiomyocytes. EE-mediated autoregulation and heterometric autoregulation probably constitute the earliest or most primitive controlling mechanisms of the heart. Coronary circulation and, later still, cardiac innervation are relatively late phylogenetic and embryological developments. These latter autoregulatory systems have become complementary to EE-mediated and heterometric autoregulation. In more compact portions of the ventricular wall (micro)vascular endothelium may be expected to become more important than EE in control of myocardial function. Reprinted with permission from Brutsaert DL, NIPS 1993; 8:82-86.

cardiomyocytes and mediate feedback through interaction with the superfusing blood. Intracavitary autoregulation by endocardial endothelium and intracoronary autoregulation by microvascular endothelium add a unique and independent type of controlling mechanism to the system.

The purpose of this chapter is twofold. First, the effects of selectively damaging cardiac endothelium on cardiac muscle dynamics and $Ca^{2+}{}_i$-transients will be reviewed in order to describe the nature and mechanisms of endothelium-mediated control of myocardial performance. Second, in order to assess the significance of endothelial control in the adaptations of the ventricular pump-function to peripheral demands, endothelium-mediated autoregulation will be compared with heterometric and homeometric autoregulation.

1. NATURE OF ENDOTHELIAL CONTROL OF MYOCARDIAL PERFORMANCE

1.1. EFFECTS OF DAMAGING ENDOCARDIAL ENDOTHELIUM ON ISOLATED CARDIAC MUSCLE PERFORMANCE

In 1988 Brutsaert et al reported for the first time that a selective damage of the *endocardial endothelium* (EE) from isolated cat papillary muscles modified the pattern of the isotonic and isometric twitch.[3] The change in the pattern of the twitch was unusual; EE-damage resulted in an immediate and irreversible abbreviation of the twitch; the onset of tension decline occurred earlier during the isometric twitch with a concomitant decrease in peak twitch tension, but without change in the early phase of the twitch. Maximal unloaded shortening velocity (Vmax) remained unchanged. This differs from virtually all other negative inotropic interventions such as decreasing extracellular calcium or reduction of cAMP-mediated effects, which are all associated with major changes in Vmax and in the early phase of the isometric twitch.

These initial observations were made on isolated cat papillary muscle at 2.5 mM $[Ca^{2+}]_o$ and 29°C. The EE was damaged by exposing the muscle stretched at lmax for 1s to 1% Triton X-100, a mild detergent dissolved in a Krebs-Ringer solution, and followed by an immediate and abundant wash. Morphological observations confirmed that this procedure sufficed to selectively damage the EE. The 1s Triton exposure was only a negligible fraction of the time needed to damage the subjacent myocardium. The effects of selective EE-damage on the mechanical performance of isolated cardiac muscle were confirmed in different animal species,[12,13,26] with various chemical,[3] physical[2] and pharmacological[18,25] damaging procedures, and at different stimulation frequencies, muscle lengths, temperatures and extracellular calcium concentrations.[3] In particular, the effect of EE-damage was more pronounced at physiological temperature (35°C) and physiological $[Ca^{2+}]_o$ (1.25 mmol/l).

Figure 3.2 characterizes and compares the effects at lmax, 35°C and 1.25 mmol/l $[Ca^{2+}]_o$ of EE-damage by Triton-immersion with two other methods (continuous wave ultrasound and air drying) which were, in preliminary experiments, elaborated to completely destroy the EE and to directly affect the underlying myocytes only minimally. Irrespective of the method, EE-destruction induced a typical early onset of relaxation, abbreviating the twitch and decreasing isometric twitch tension, as described earlier. The effect on the onset of tension decline and duration of the twitch was not different for the three procedures (time from stimulus to half isometric relaxation decreased by ± 10%). At 35°C, Triton affected the early phase of the isometric twitch substantially; maximal rate of tension development decreased by more than 30%. This contrasted to the effects of ultrasound and of air drying and to the initial observations with Triton at 29°C. Regardless of the method and temperature, however, peak isometric tension and maximal rate of tension development were unaltered at high $[Ca^{2+}]_o$ (Fig. 3.2). These observations, together with confocal or electron microscopic observations, excluded functional and/or structural destruction of the subjacent myocardium by Triton, ultrasound or air drying.

Quantification of the Effect of a Complete EE-Destruction on Cardiac Muscle Performance is Important

1. Quantification is needed to assess the significance of EE-mediated regulation in comparison with the traditional length-mediated and neurohumoral-mediated regulation of cardiac muscle performance.

2. Quantification may be useful in the elucidation of the signaling mechanisms between EE and the underlying myocardium. Several investigators have tried to reverse the effects of EE-removal on twitch contraction by exogenous administration of EE-cell superfusate[28] or of endothelium-derived substances such as endothelin,[29] and compared the effect of EE-removal with the effect of blockers of endothelial-derived substances.[7] Although these experiments already gave some insight in the possible mechanisms of EE-mediated myocardial control, these experiments should, in our opinion, be performed by using different damaging methodologies at physiological conditions before definitive conclusions about the signaling mechanisms between the EE and the myocardium can be made.

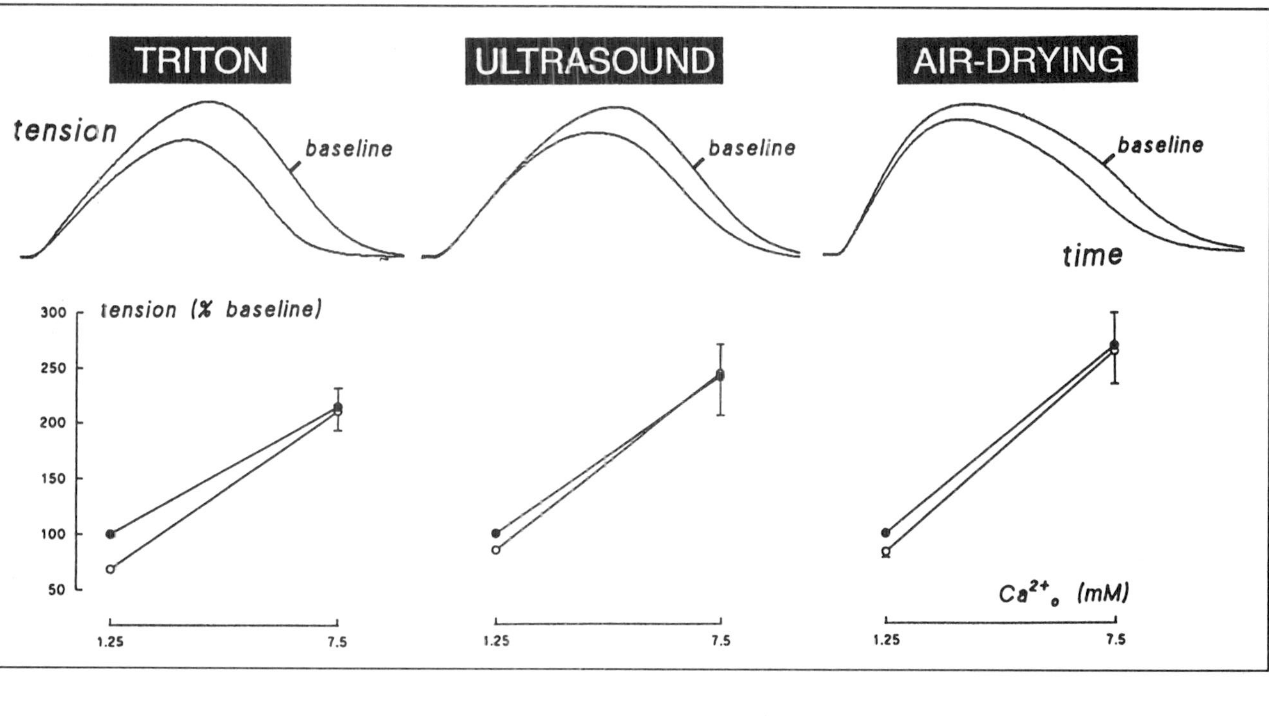

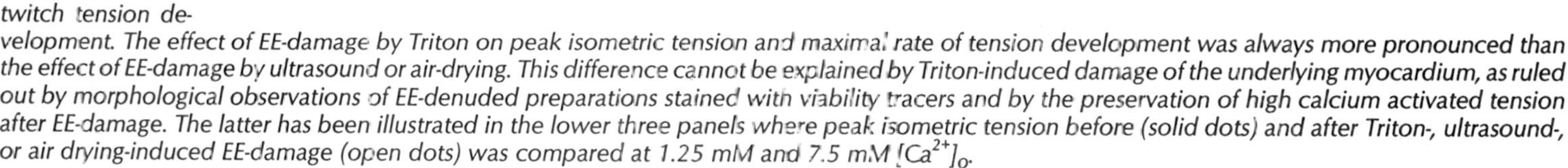

Fig. 3.2. Contractile effects of damaging endocardial endothelium (EE) by Triton X-100 (left), ultrasound (middle), or air-drying (right). The three upper panels show isometric twitches before (baseline) and after EE-damage (at l_{max}, 35 °C and 1.25 mM Ca^{2+}). Irrespective of how EE was damaged, the profile of the change in pattern of the twitch was similar, i.e. shortening of the duration of the twitch with concomitant decrease in peak isometric twitch tension and late rise in twitch tension development. The effect of EE-damage by Triton on peak isometric tension and maximal rate of tension development was always more pronounced than the effect of EE-damage by ultrasound or air-drying. This difference cannot be explained by Triton-induced damage of the underlying myocardium, as ruled out by morphological observations of EE-denuded preparations stained with viability tracers and by the preservation of high calcium activated tension after EE-damage. The latter has been illustrated in the lower three panels where peak isometric tension before (solid dots) and after Triton-, ultrasound-, or air drying-induced EE-damage (open dots) was compared at 1.25 mM and 7.5 mM $[Ca^{2+}]_o$.

1.2 Effects of Damaging Endocardial Endothelium on Intact Ventricular Performance

Removal of the EE in isolated Langendorff perfused ferret heart[8] by transient exposure to Triton X-100 shortened the duration of contraction as it does in the isolated cardiac muscle preparation. The development of a catheter-mounted, high power, high frequency, continuous wave ultrasound probe allowed Gillebert et al and De Hert et al in 1991 to extend the above in vitro observations to in vivo conditions.[6,9] In analogy with the results in isolated muscle, partial and selective EE damage by intracavitary ultrasound irradiation of the ventricular wall of the heart of anesthetized open chest dogs, resulted in premature pressure fall during ventricular relaxation (Fig. 3.4); the ensuing shortened systole was accompanied by a pronounced early reextension of the myocardial fibers in several parts of the ventricular wall. These alterations in systolic duration and onset of rapid filling in the intact dog, as in isolated muscle, likewise amounted up to about 10% abbreviation in the timing of pressure fall (decrease in time to -dP/dt). Accordingly, the EE appears to exert a tonic effect on left ventricular contraction also in the whole heart.

1.3. Effects of Damaging (Micro)Vascular Endothelium on Isolated Cardiac Muscle Performance

We have so far considered a modulatory role on cardiac contractile behavior as mediated by endothelial cells from the endocardium. The heart contains also large quantities of vascular endothelial cells in the coronary microvasculature in immediate proximity to the myocardium. There is now compelling evidence that coronary (micro)vascular endothelial cells may *directly* affect contractile performance of the subjacent cardiomyocytes.[13,14,22] At physiological calcium, the baseline isometric twitch characteristics of isolated rabbit papillary muscles with dysfunctional vascular endothelium were modified in a manner typical of that created by EE-dysfunction; twitches were significantly shorter but there was no marked difference in early contraction dynamics when compared with twitches of muscles with intact (micro)vascular endothelium. EE removal in these muscles caused similar changes on twitch characteristics whether (micro)vascular endothelium had been made dysfunctional or not. Accordingly, coronary vascular endothelium directly modulates the contractile characteristics of adjacent myocardium in a manner similar and additive to the modulation by the EE.

In conclusion, cardiac endothelium both in vitro and in vivo directly modulates or controls the performance of the subjacent cardiomyocytes. As selective damage of cardiac endothelium leads to decrease in twitch amplitude and duration, the presence of intact cardiac endothelium imparts a positive inotropic and contraction prolonging effect on the myocardium.

2. MECHANISMS OF ENDOTHELIAL CONTROL OF MYOCARDIAL PERFORMANCE

Inotropic stimuli alter contractile muscle performance by modulating the $[Ca^{2+}]_i$-transient ("Ca^{2+}-availability") or the myofibrillar responsiveness to $[Ca^{2+}]_i$ (Ca^{2+}-sensitivity). Each parameter of muscle twitch, influenced by inotropic stimuli, must therefore represent a change in Ca^{2+}-availability or Ca^{2+}-sensitivity or both. Because of the similarity in the pattern of changes induced by the presence or absence of a functional EE and those induced by increasing initial muscle length in vitro or end-diastolic volume in vivo (Fig. 3.4), one might hypothesize that EE-dependent events may be mediated through similar mechanisms as length-dependent events, i.e. through increased responsiveness of the myofilaments to $[Ca^{2+}]_i$.[3]

This hypothesis has been endorsed by several experimental observations. Wang and Morgan examined the effect of EE-damage (by Triton, 30°C) on $[Ca^{2+}]_i$-transient and peak isometric twitch of ferret papillary muscles loaded with the Ca^{2+}-regulated bioluminescent indicator aequorin (Fig. 3.3, left panel).[29] These experiments supported the hypothesis that the EE modulates contractile performance through responsiveness of the contractile proteins. Interestingly, EE-damage was accompanied by an increase in peak $[Ca^{2+}]_i$ (but not resting $[Ca^{2+}]_i$), even in presence of ryanodine and especially at the higher $[Ca^{2+}]_o$.

The use of the relation between peak $[Ca^{2+}]_i$ and associated peak isometric twitch tension has, however, been severely criticized,[10] essentially because peak $[Ca^{2+}]_i$ and peak tension are derived from transient (i.e. not steady state) phenomena and are temporally dissociated. The putative modulatory properties of EE on the contractile proteins have, nevertheless, been endorsed further by various other experimental approaches.

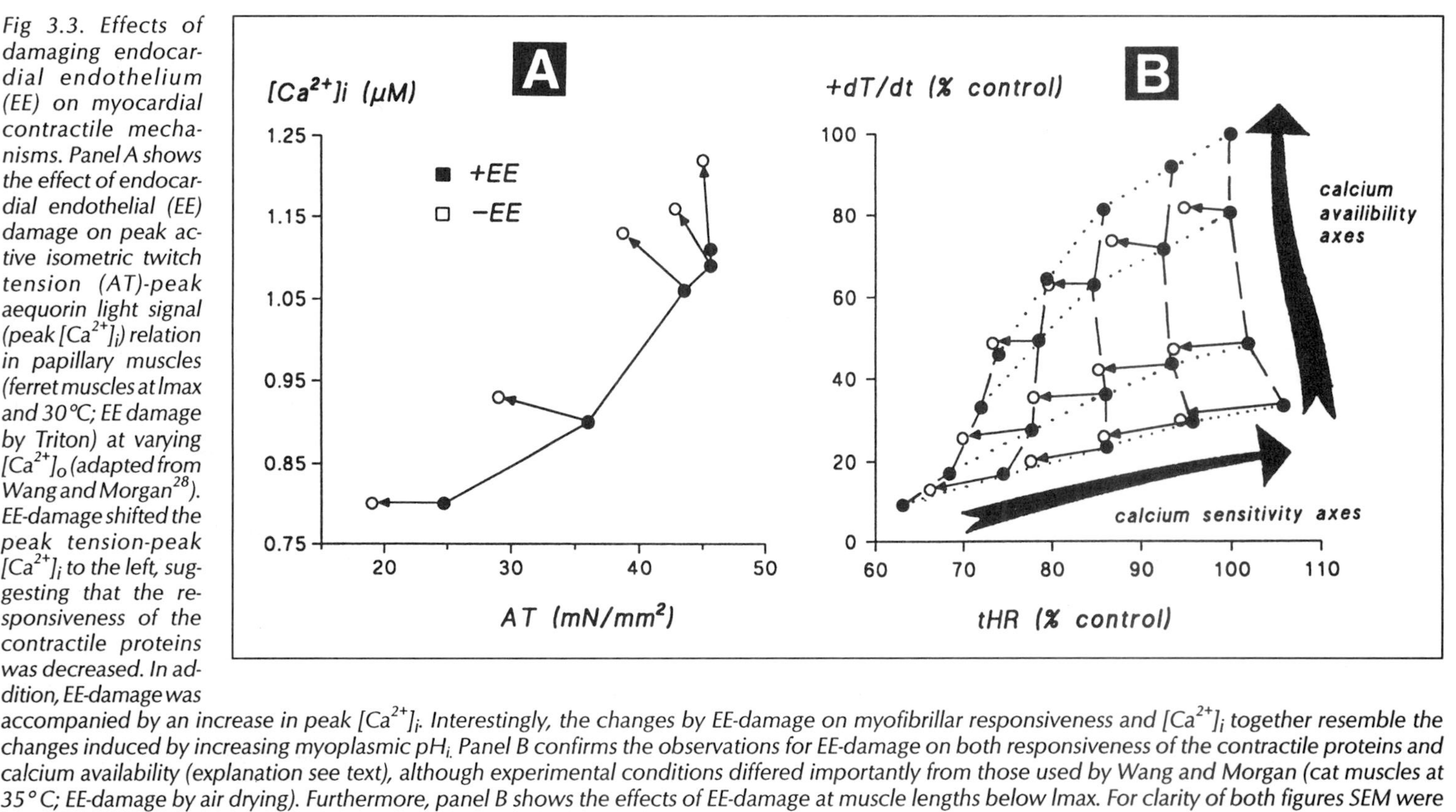

Fig 3.3. Effects of damaging endocardial endothelium (EE) on myocardial contractile mechanisms. Panel A shows the effect of endocardial endothelial (EE) damage on peak active isometric twitch tension (AT)-peak aequorin light signal (peak $[Ca^{2+}]_i$) relation in papillary muscles (ferret muscles at lmax and 30°C; EE damage by Triton) at varying $[Ca^{2+}]_o$ (adapted from Wang and Morgan[28]). EE-damage shifted the peak tension-peak $[Ca^{2+}]_i$ to the left, suggesting that the responsiveness of the contractile proteins was decreased. In addition, EE-damage was accompanied by an increase in peak $[Ca^{2+}]_i$. Interestingly, the changes by EE-damage on myofibrillar responsiveness and $[Ca^{2+}]_i$ together resemble the changes induced by increasing myoplasmic pH_i. Panel B confirms the observations for EE-damage on both responsiveness of the contractile proteins and calcium availability (explanation see text), although experimental conditions differed importantly from those used by Wang and Morgan (cat muscles at 35° C; EE-damage by air drying). Furthermore, panel B shows the effects of EE-damage at muscle lengths below lmax. For clarity of both figures SEM were not included.

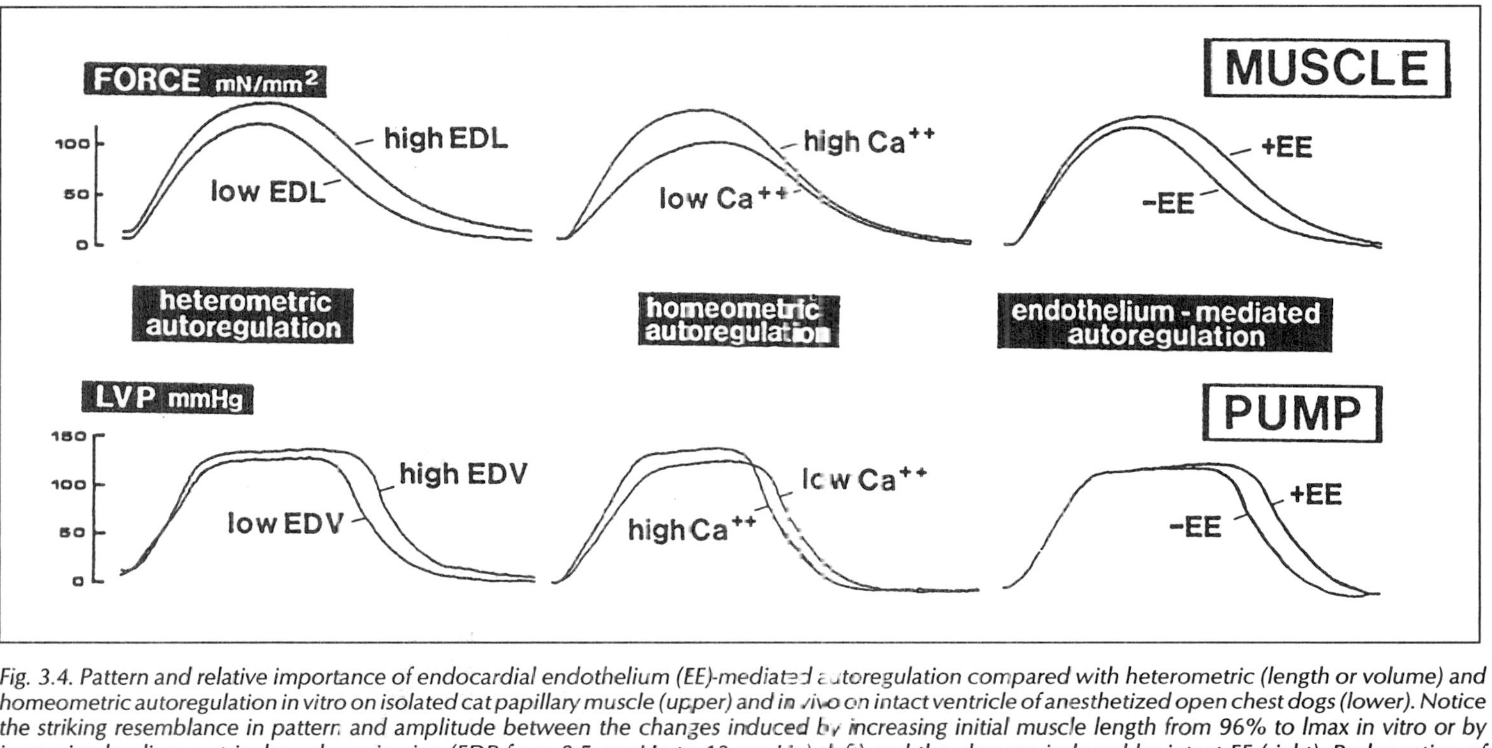

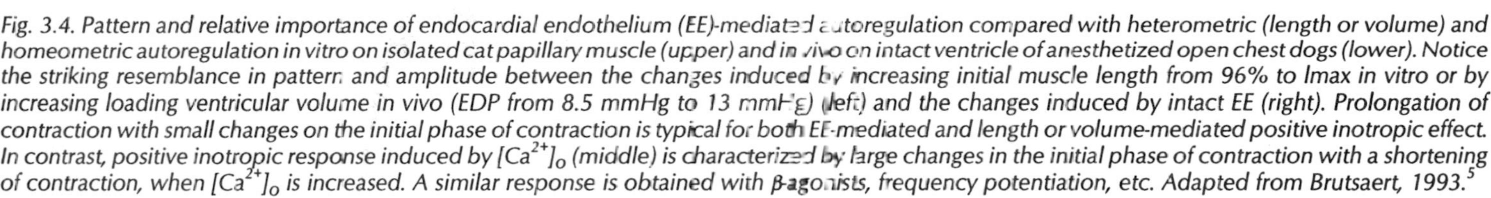

Fig. 3.4. Pattern and relative importance of endocardial endothelium (EE)-mediated autoregulation compared with heterometric (length or volume) and homeometric autoregulation in vitro on isolated cat papillary muscle (upper) and in vivo on intact ventricle of anesthetized open chest dogs (lower). Notice the striking resemblance in pattern and amplitude between the changes induced by increasing initial muscle length from 96% to lmax in vitro or by increasing loading ventricular volume in vivo (EDP from 8.5 mmHg to 13 mmHg) (left) and the changes induced by intact EE (right). Prolongation of contraction with small changes on the initial phase of contraction is typical for both EE-mediated and length or volume-mediated positive inotropic effect. In contrast, positive inotropic response induced by $[Ca^{2+}]_o$ (middle) is characterized by large changes in the initial phase of contraction with a shortening of contraction, when $[Ca^{2+}]_o$ is increased. A similar response is obtained with β-agonists, frequency potentiation, etc. Adapted from Brutsaert, 1993.[5]

First, Wang and Morgan extended their observations to steady state $[Ca^{2+}]_i$ and tension derived from tetanic contractions from ferret papillary muscles preincubated with ryanodine.[29]

The *second* line of evidence has been recently elaborated in our laboratory and arose from a novel analysis of isolated muscle mechanics. This analysis unmasks cellular mechanisms underlying changes in inotropic state and overcomes problems and limitations inherent to experimental methodologies such as aequorin or pharmaceutical interventions such as ryanodine. Before and after an inotropic intervention isometric twitch parameters are measured at different degrees of Ca^{2+}-sensitivity (obtained by sudden changes in initial muscle length) and Ca^{2+}-availability (induced by changes in $[Ca^{2+}]_o$). In the isometric twitch, variations in Ca^{2+}-sensitivity are manifested mainly by changes in the duration of the twitch (exemplified by the time from stimulus to half isometric relaxation; THR), whereas variations in Ca^{2+}-availability are mainly manifested by changes in the rate of tension development (exemplified by peak rate of tension development +dT/dt). The diagram in Figure 3.3 (right panel) plots THR to +dT/dt at different muscle lengths and $[Ca^{2+}]_o$ for a whole muscle group (n = 7; mean values). The broken lines exemplify changes induced by variations in $[Ca^{2+}]_o$ at constant muscle length and thus stand for Ca^{2+}-availability axes. The dotted lines exemplify changes induced by variations in length at constant $[Ca^{2+}]_o$ and thus stand for Ca^{2+}-sensitivity axes. Each inotropic intervention imposed on the muscle induces a shift along the Ca^{2+}-sensitivity axis, or the Ca^{2+}-availability axis or both. As can be noticed from Figure 3.3, the negative inotropic effect induced by EE-damage (air drying, 35°C), at 1.25 mmol/l $[Ca^{2+}]_o$, was characterized by a leftward shift along the Ca^{2+}-sensitivity axis. At the higher $[Ca^{2+}]_o$, the shift was accompanied by an increase in Ca^{2+}-availability. These results thus reveal that EE-damage from an isolated cardiac muscle causes a decrease in Ca^{2+}-sensitivity in the myocardium and, especially at the higher $[Ca^{2+}]_o$, an increase in Ca^{2+}-availability. The resemblance of these results with the results of Wang and Morgan with the aequorin method are striking.

Finally, several recent experiments have demonstrated that EE-derived substances, in particular endothelin and nitric oxide, may influence the sensitivity of the contractile proteins, respectively by altering myoplasmic pH_i through activation of the Na/H^+ exchanger[11] or by cGMP-mediated processes.[27]

In conclusion, there is compelling experimental evidence that the presence of intact EE enhances myocardial contractile performance through increasing the responsiveness of the contractile proteins to $[Ca^{2+}]_i$. In addition, experiments suggest that this phenomenon is associated with a change in $[Ca^{2+}]_i$ in the opposite direction. Interestingly, the modulatory effects of EE resemble changes in contractile protein responsiveness and on intracellular Ca^{2+}-availability induced by changes in myoplasmic pH_i. Some of the EE-derived factors, such as endothelin, have indeed been shown to alter pH_i. Endothelium-mediated control of myoplasmic pH_i is an intriguing hypothesis which could explain many of the above observations, although it needs further investigation.

3. ENDOTHELIUM-MEDIATED AUTOREGULATION VERSUS HETEROMETRIC AND HOMEOMETRIC AUTOREGULATION

We have seen above that endothelial cells from both the endocardium and the coronary (micro)vasculature modulate the contractile performance of the underlying cardiomyocytes. From these observations, we have postulated the existence of endothelium-mediated autoregulation of cardiac performance. Intracavitary autoregulation by endocardial endothelium and intracoronary autoregulation by microvascular endothelium may add a unique type of controlling system of the heart, where endothelial cells sense homeostatic variations in the superfusing blood, transmit them to the underlying myocardium and participate in regional and global adjustments of the subjacent myocardium to these variations.

The relative contribution of both types of regulation is more difficult to predict. Although the coronary endothelium may be considered by some to be a more likely candidate for modulation of the myocardium than the EE, the functional roles of EE and coronary endothelium were shown to be additive[13] and may be supposed to be complementary if one takes into account the following considerations. First, their relative contribution will depend on the ventricular surface-to volume ratio or the relative amount of EE cells and of coronary vascular endothelial cells per unit of weight of cardiomyocytes in any given region of the cardiac wall. On the other hand, the endothelial surface area, which has been shown to be immense for EE and is necessary for sensory function, has to be considered. Therefore, on theoretical grounds, endothelial cells in the coronary system might be expected to exert a local control, mainly responding to changes in local blood flow and shear stress, whereas the sensory function of the EE might perhaps be attributed a more global role, whereby substances in the blood circulating through the ventricles could induce immediate changes in overall myocardial performance.

Accumulating the data confirming the concept of endothelium-mediated autoregulation, one cannot get around the question how endothelium-mediated control relates to other control mechanisms of cardiac performance, e.g., through volume loading (Starling; heterometric control) or through neurohumoral activation (homeometric autoregulation) (Fig. 3.4).

When left ventricular (LV) end-diastolic pressure was elevated from low (4.1 ± 0.9 mmHg) to baseline (10.6 ± 1.5 mmHg) or from baseline to high (17.9 ± 1.8 mmHg) in anesthetized open chest dogs, time to peak -dP/dt increased with a mean value of 12.2% and 4.3% respectively (Fig. 3.4).[5] Comparison of these variations in systolic duration over reasonably wide ranges of heterometric autoregulation with the variations by the above range of EE-mediated autoregulation in vivo suggests that both types of autoregulation may be nearly as powerful in day-to-day adaptations of cardiac performance. Interestingly, from these considerations it would seem that the two phylogenetically and embryologically most primitive autoregulatory control systems act over a similar range and through variations in the responsiveness of the contractile proteins to $[Ca^{2+}]_i$. These variations are manifested mechanically as changes in the onset of relaxation and the duration of systole, with or without changes in the rate of relaxation. Increased sensitivity does not necessarily imply a slower relaxation, since the direction of the change in rate of relaxation would be determined by the precise mechanism of action (on- or off- rate constant of Ca^{2+}-troponin C or actin-myosin interaction).

The hemodynamic consequences of altering the onset of relaxation and of systolic duration are, however, only beginning to be explored. Modulation of the onset of relaxation will influence early diastolic filling. Since filling of the heart is as important as ejection, altered onset of relaxation may have important consequences in terms of pump function, especially in conditions of impaired relaxation.

A quantitative comparison between neurohumoral (homeometric) control and endothelium-mediated control is difficult, as the net effect on cardiac function of neurohumoral control is the result of impact on cardiac and peripheral system by a variety of hormones and transmitters, the individual concentrations of which may fluctuate continuously. Several in vitro studies have shown that the EE participates in the inotropic response to several of these substances. The in vitro inotropic response to atrial natriuretic peptide[17] and to a low dose of the α_1-agonist phenylephrine[18] required the presence of an intact EE. For other substances, e.g., serotonin,[25] vasopressin,[23] acetylcholine[20] and angiotensin (I and II),[19] the EE was shown to modify the inotropic response. At high levels of epinephrine, vasopressin, serotonin or atrial natriuretic peptide, the EE was selectively damaged. Hence, there seems to be in vitro evidence to postulate that the EE mediates the effect of major extrinsic (homeometric) cardiac compensatory mechanisms. Because in chronic heart failure plasma levels of most of the above substances are increased, one wonders to what extent elevated plasma levels could create a point of no return in the pathogenesis of the disease, possibly by converging to some state of irreversible dysfunction of the EE.[4]

4. GENERAL CONCLUSION

Cardiac endothelium in the endocardium and coronary (micro)vasculature directly influence the performance of the myocardium. In nonstimulated conditions, cardiac endothelium imparts a positive inotropic effect on isolated cardiac muscle and intact ventricle by increasing the responsiveness of the contractile proteins. Cardiac endothelium-mediated control of myocardial performance resembles length/volume-mediated control and interacts, through its sensory function, with neurohumoral-mediated control. Cardiac endothelium seems thus to be involved in adapting cardiac performance to peripheral demands. It is, however, not yet fully established how cardiac endothelial function is regulated and to what extent integrity of cardiac endothelium control of the myocardium is important for normal function of the heart.

Two major possible mechanisms have been postulated to explain cell-to-cell communication between the endocardial endothelial cell and the immediately subjacent cardiomyocyte, i.e. either through an active electromechanical blood-heart barrier function or through a stimulus secretion coupling. Both mechanisms, which may either act together or in parallel, will be dealt with in chapters 4 and 5 respectively.

References

1. Allen DG, Kurihara S. The effects of muscle lengths on intracellular calcium transient in mammalian cardiac muscle. J Physiol (London)1982; 327:79-94.
2. Andries LJ, Meulemans AL, Brutsaert DL. Ultrasound as a novel method for selective damage of the endocardial endothelium. Am J Physiol 1991, 261:H1636-H1642.
3. Brutsaert DL, Meulemans AL, Sipido KR et al. Effects of damaging the endocardial surface on the mechanical performance of isolated cardiac muscle. Circ Res 1988; 62:357-366.
4. Brutsaert DL. Role of endocardium in cardiac overloading and failure. Eur Heart J 1991; 11:8-16.
5. Brutsaert DL. Endocardial and coronary endothelial control of cardiac performance. NIPS 1993; 8:82-86.
6. De Hert SG and Gillebert TC. Alterations of left ventricular endocardial function by intracavitary high power ultrasound interacts with volume, inotropic state and α1-adrenergic stimulation. Circ 1993; 87:1275-1285.
7. Evans HG, Lewis MJ, Shah AM. Modulation of myocardial relaxation by basal release of endothelin from endocardial endothelium. Cardiov Res 1994; 28:1694-1699.
8. Fort S, Lewis MJ and Shah AM. The role of endocardial endothelium in the modulation of myocardial contraction in the isolated whole heart: Cardioscience 1993; 4:217-222.
9. Gillebert TC, De Hert SG, Andries LJ et al. Intracavitary ultrasound impairs left ventricular performance: presumed role of endocardial endothelium. Am J Physiol 1992,263:H857-H865.
10. Gwathmey JK and Hajjar RJ. Relation between steady-state force and intracellular [Ca^{2+}] in intact human myocardium. Index of myofibrillar responsiveness to Ca^{2+}. Circ 1990; 82:1266-1278.
11. Kramer BK, Smith TW, Kelly RA. Endothelin and increased contractility in adult rat ventricular myocytes: role of intracellular alkalosis induced by activation of the protein kinase C-dependent Na^+-H^+-exchanger. Circ Res 1991; 68:269-279.
12. Li K, Rouleau JL, Calderone A et al. Endocardial function in pacing-induced heart failure in the dog. J Moll Cell Cardiol 1993; 25:529-538.
13. Li K, Rouleau JL, Andries LJ et al. Effect of dysfunctional vascular endothelium on myocardial performance in isolated papillary muscles. Circ Res 1993; 72:768-777.
14. McClellan G, Weisberg A, Kato NS et al. Contractile proteins in myocardial cells are regulated by factors released by blood vessels. Circ Res 1992; 70:787-803.
15. Mebazaa A, Mayoux E, Maeda K et al. Paracrine effects of endocardial endothelial cells on myocyte contraction via endothelin. Am J Physiol 1993; 265:H1841-H1846.
16. Mebazaa A, Martin LD, Robotham JL et al. Right and left ven-

tricular cultured endocardial endothelium produces prostacyclin and PGE_2. J Moll Cell Cardiol 1993; 25:245-248.
17. Meulemans AL, Andries LJ and Brutsaert DL. Endocardial endothelium mediates positive inotropic response to α1-adrenoreceptor agonist in mammalian heart. J Moll Cell Cardiol 1990; 20:667-685.
18. Meulemans AL, Sipido KR, Sys SU et al. Atriopeptin III induces early relaxation of isolated mammalian papillary muscle. Circ Res 1988; 62:1171-1174.
19. Meulemans AL, Andries LJ and Brutsaert DL. Does endocardial endothelium mediate positive inotropic response to angiotensin I and angiotensin II? Circ Res 1990; 66:1591-1601.
20. Mohan P, Brutsaert DL and Sys SU. Positive inotropic effect of acetylcholine; role of endocardial endothelium, cGMP, nitic oxide, prostaglandins. Circ 1994; 90:I649.
21. Orchard CH and Kentish JC. Effects of changes of pH on the contractile function of cardiac muscle. Am J Physiol 1990; 258:C967-C981.
22. Ramaciotti C, Sharkey A, Mc Clellan G et al. Endothelial cells regulate cardiac contractility. Proc Natl Acad Sci USA 1992; 89:4033-4036.
23. Schoemaker IE, Meulemans AL, Andries LJ et al. Role of the endocardial endothelium in the positive inotropic action of vasopressin. Am J Physiol 1990; 259:H1148-H1151.
24. Schulz R, Smith JA, Lewis MJ et al. Nitric oxide synthase in cultured endocardial cells of the pig. Br J Pharmacol 1991; 104:21-24.
25. Shah AM, Andries LJ, Meulemans AL et al. Endocardium mediates inotropic response to 5-hydroxytryptamine. Am J Physiol 1989; 257:H1790-H1797.
26. Shah AM, Smith JA and Lewis MJ. The role of endocardium in the modulation of contraction of isolated papillary muscle of the ferret. J Cardiov Pharmacol 1991; 17(S3):S251-S257.
27. Shah AM, Spurgeon HA, Sollot SJ et al. 8-bromo cGMP reduces the myofilament response to calcium in intact cardiac myocytes. Circ Res 1994; 74:970-978.
28. Smith JA, Shah AM and Lewis MJ. Factors released from endocardium of the ferret and pig modulate myocardial contraction. J Physiol (London) 1991; 439:1-14.
29. Wang J and Morgan JP. Endocardial endothelium modulates myofilament Ca^{2+}-responsiveness in aequorin-loaded ferret myocardium. Circ Res 1992; 70:754-760.
30. Yue DT. Intracellular $[Ca^{2+}]$ related to rate of force development in twitch contraction of the heart. Am J Physiol 1987; 352:H760-H770.

CHAPTER 4

Physicochemical Properties of the Endocardial Endothelium

Paul Fransen and Marc Demolder

The endocardial endothelium (EE) forms a continuous monolayer of closely apposed cells covering the complete lumen of the cardiac cavities (chapter 2). It has been demonstrated to be an important modulator of cardiac performance (chapter 3). How signals from EE are transduced to the underlying myocytes is still under investigation (chapter 5). Two mechanisms of signal transduction have been proposed.[5] On the one hand, EE cells may release endothelium-derived factors which alter the contractile state of the cardiomyocytes (stimulus-secretion-contraction coupling). On the other hand, the EE may act as a physicochemical barrier, controlling the specific ionic constitution of the interstitial milieu of the heart muscle cells and, thereby, modulating cardiac performance (blood-heart barrier).

The aim of the present chapter is to describe basic electrophysiological properties of EE cells and to elucidate the role of ion channels, pumps and transporters in both signal transduction mechanisms.

1. STIMULUS-SECRETION-CONTRACTION COUPLING

Vascular and endocardial endothelial cells respond to physical and humoral stimuli by secreting a number of biologically active

Endocardial Endothelium: Control of Cardiac Performance, edited by Stanislas U. Sys and Dirk L. Brutsaert.

substances. Some of these endothelium-derived substances like endothelium-derived relaxing factor (EDRF or NO), prostacyclin (PGI_2) and endothelin have an influence on excitation-contraction coupling in smooth muscle cells and cardiomyocytes (for references, see chapter 5). Thereby, they modulate vascular tone and myocardial performance. A crucial step in the stimulus-secretion coupling at the level of the endothelial cells is the increase in the intra-endothelial Ca^{2+}-concentration ($[Ca^{2+}]_i$). In both vascular and endocardial endothelial cells, the increase in $[Ca^{2+}]_i$ after physical or humoral stimulation occurs in two phases.[3] After an initial transient increase due to the release of Ca^{2+}-ions from the intracellular stores, $[Ca^{2+}]_i$ remains elevated because of an influx of Ca^{2+}-ions from the extracellular space (Fig. 4.1A[23]). The influx of Ca^{2+} from the extracellular space was abolished after depolarization of the EE cells (Fig. 4.1B), indicating that the resting membrane potential is an important parameter, which affects endothelial Ca^{2+}-homeostasis and release of endothelium-derived factors.[4,21,22,23]

1.1. Resting Membrane Potential of Endocardial Endothelial Cells

The resting membrane potential (V_m) has been determined in freshly isolated and in cultured endocardial endothelial cells from different species (Table 4.1).

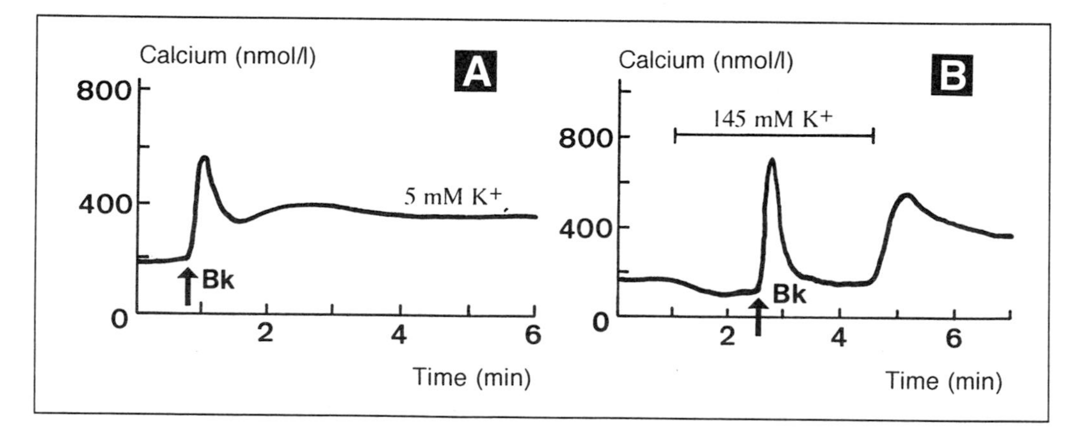

Fig. 4.1. Endocardial endothelial monolayers (cultured bovine atrial endothelial cells) exhibit a biphasic $[Ca^{2+}]_i$ response to bradykinin (200 nmol/l) stimulation in normal physiological saline (5 mmol/l $[K^+]_o$) (A). Under depolarizing conditions (equimolar replacement of $[Na^+]_o$ by $[K^+]_o$) the basal $[Ca^{2+}]_i$ is depressed and the $[Ca^{2+}]_o$-dependent plateau phase is abolished in a reversible manner (B). Modified after Laskey at al.[23]

Table 4.1. Resting membrane potentials of non-stimulated EE cells from different species and studied in different conditions

EE Cell Preparation	V_m (mV)	References
cultured bovine atrial cells 5 mM $[K^+]_o$, 22°C	-66.6 ± 2.1 (SEM, n = 16)	Laskey et al[21,22]
guinea pig EE cells		Manabe et al[25]
intact small tissue blocks	-44.4 ± 0.8 (SEM, n = 118)	
freshly isolated cells	-42.6 ± 3.0 (SEM, n = 8)	
cultured porcine right ventricle cells, 5.4 mmol/l $[K^+]_o$, 37°C	-66.5 ± 8.5 (SD, n = 35)	Fransen et al[12,13]

$[K^+]_o$ represents the extracellular K^+-concentration.

Values for V_m in resting EE cells were comparable with values reported in intact arterial endothelium and isolated cultured macrovascular endothelial cells (V_m between -41 and -77 mV[1]). Similarly as in vascular endothelial cells, the negative value of V_m in endocardial endothelial cells was dependent on $[K^+]_o$, indicating that in nonstimulated EE cells the membrane is mainly permeable to K^+-ions.[13,21,22]

In nonstimulated cultured EE cells of the porcine right ventricle, the main membrane current under whole-cell voltage-clamp conditions was the inwardly rectifying K^+-current (I_{Ki}) (Fig. 4.2). In 70% of the cells, outwardly directed currents at positive potentials were small (< 2 pA/pF) as in the cell of Figure 4.2. In 30% of the cells, an outwardly rectifying current (> 2 pA/pF at +60 mV) was observed. In the former group of cells, the current-voltage (I-V) relation approached zero-currents between V_m and +60 mV. Within this range of potentials, small changes in membrane current (current-clamp-mode) induced large changes in the value of the resting potential of the cells from this group. During measurement of the zero-current (or resting) potential small changes in membrane current can be due to leakage currents between the pipette tip and the cell membrane.[6] In six cells, the zero-current potential was determined without and after correction for the seal leakage currents and the zero-current potential shifted from -56 ± 11(SD) mV to -71 ± 4 (SD) mV. So, a variable seal leakage

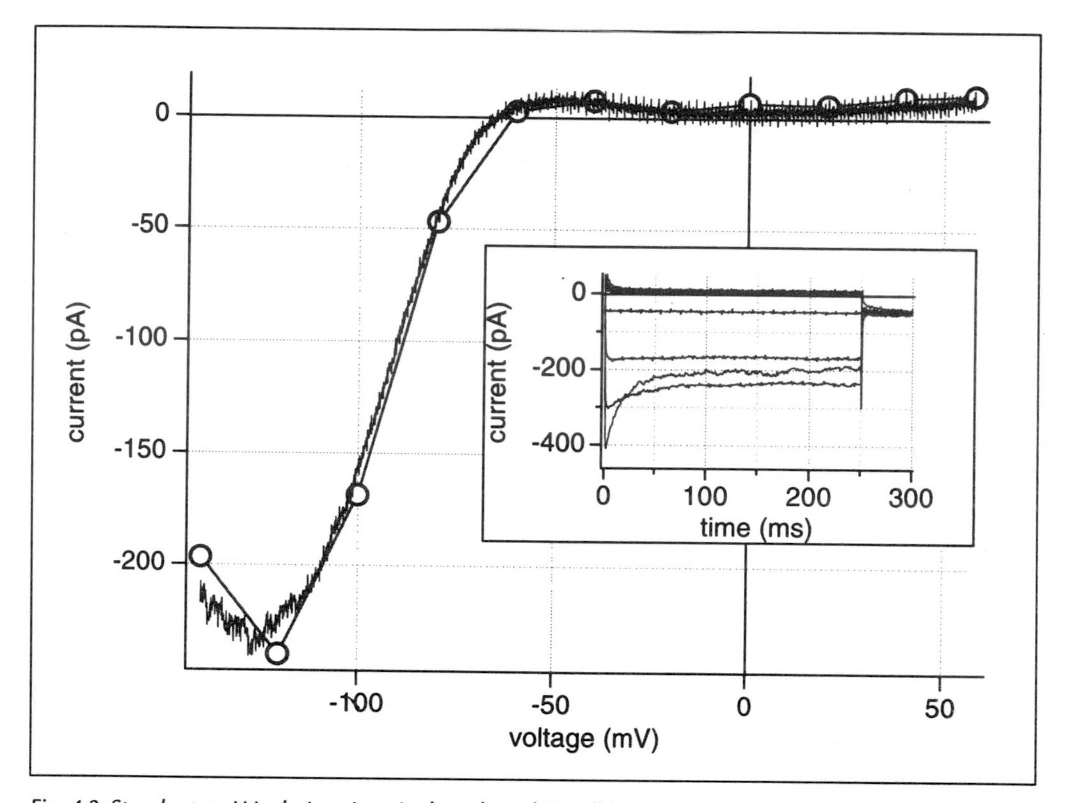

Fig. 4.2. Steady-state I-V relations in a single, cultured EE cell from the porcine right ventricle. The bathing solution contained (in mmol/l): 5.4 KCl, 128 NaCl, 1.8 $CaCl_2$, 0.8 $MgCl_2$, 10 HEPES, 10 glucose, 300-310 mOsm, pH = 7.4, 37°C. The patch pipette was filled with (in mmol/l): 30 KCl, 100 K-aspartate, 10 NaCl, 1 $CaCl_2$, 10 EGTA, 5 $MgCl_2$, 10 HEPES, 300-310 mOsm, pH = 7.2, 37°C, resistance 2-5 MΩ. The conventional whole-cell mode of the patch-clamp technique was used to apply 11 consecutive clamps in 20 mV steps from -140 to +60 mV from the holding potential of -80 mV. The corresponding membrane currents as a function of clamp time (250 ms) are shown in the inset. At -140 and -120 mV, currents were time-dependent and decreased with time. This apparent inactivation of I_{Ki}-currents is due to block by external Na^+-ions. Outwardly directed currents at positive potentials were small and not time-dependent. The steady-state I-V relation of these currents is represented by the squares. The main current in this cell (capacity 27.4 pF, series resistance 22.6 MΩ) was the inwardly rectifying K^+-current, I_{Ki}. Superimposed on this I-V relation, there is the membrane current obtained by a ramp clamp of 100 mV/s over the same voltage range and in the same cell. The ramp clamp I-V relation corresponded very well with the I-V relation obtained with the step clamp protocol. The zero-current potential of this cell (and resting potential) was -63 mV.

current in different experiments can indeed cause large variations in the measured resting potential. Nevertheless, after correction for seal leakage currents, the zero-current potential of -71.0 ± 4.0mV (SD, n = 6) or -66.5 ± 8.5 mV (SD, n = 35, Table 4.1) was still 15 to 20 mV more depolarized than the equilibrium potential for K^+-ions (-85 mV), indicating that other ionic conductances are involved in determining the resting potential of EE cells.

An active Na^+/K^+ ATPase in EE cells should also contribute in setting the negative value of V_m. The presence of the pump in EE cells has been suggested by Laskey et al.[22] Upon removal of extracellular K^+ (replacement with Na^+), EE cells depolarized by 15 to 20 mV from -67 mV to -55/-45 mV and this depolarization was attributed to the inhibition of the Na^+/K^+ pump in zero $[K^+]_o$. Similar, but smaller depolarizations (5 to 8 mV) have been described in guinea pig coronary endothelial cells by Daut et al.[11] In experiments on cultured porcine EE cells, removal of K^+_o (Fig. 4.3A) caused a sudden and large depolarization to potentials between -40 and -20 mV. In our opinion, these large depolarizations upon inhibition of the Na/K pump cannot only be attributed to the removal of a hyperpolarizing Na^+/K^+ ATPase-current. Blocking I_{Ki} with 100 µmol/l Ba^{2+} (Fig. 4.3B) also caused large depolarizations of V_m without inhibition of the pump current. Therefore, the large depolarizations are believed to be caused by the decrease in conductance of the I_{Ki}-current by low $[K^+]_o$ or Ba^{2+} and the concomitant increase of the relative importance of other ionic conductances in determining the zero-current potential.

The contribution of ions other than K^+-ions in determining the resting potential of nonstimulated EE cells, was investigated by studying the effects of alterations in Na^+-, Ca^{2+}- and Cl^--concentrations. Replacement of extracellular Na^+ with N-methyl-D-glucamine$^+$ depolarized the resting potential of single cultured EE cells from the porcine right ventricle by 5 to 10 mV (see also Laskey et al,[22] Na^+ replacement with Li^+). In the absence of Na^+, I_{Ki}-currents became time-independent at negative potentials of -140 or -120 mV. This has also been observed in vascular endothelial cells and suggests that external Na^+ blocks I_{Ki}-channels at negative potentials.[43] Removal of extracellular Ca^{2+} from the bathing solution did not influence the reversal potential of the I-V relations in EE cells, but caused an increase in I_{Ki}-currents at all potentials, while an increase in $[Ca^{2+}]_o$ had the reverse effect.

In vascular endothelial cells, the resting potential was not influenced by changes in $[Cl^-]_o$.[2,30] In single cultured EE cells of the pig right ventricle, however, V_m or the zero-current potential was not only dependent on K^+-ions but also on Cl^--ions. Figures 4.4A and B illustrate an experiment in which the intracellular Cl^--concentration ($[Cl^-]_i$) was changed by pipette and cell perfusion. $[Cl^-]_i$ was decreased from 22 to 5 mmol/l and subsequently increased

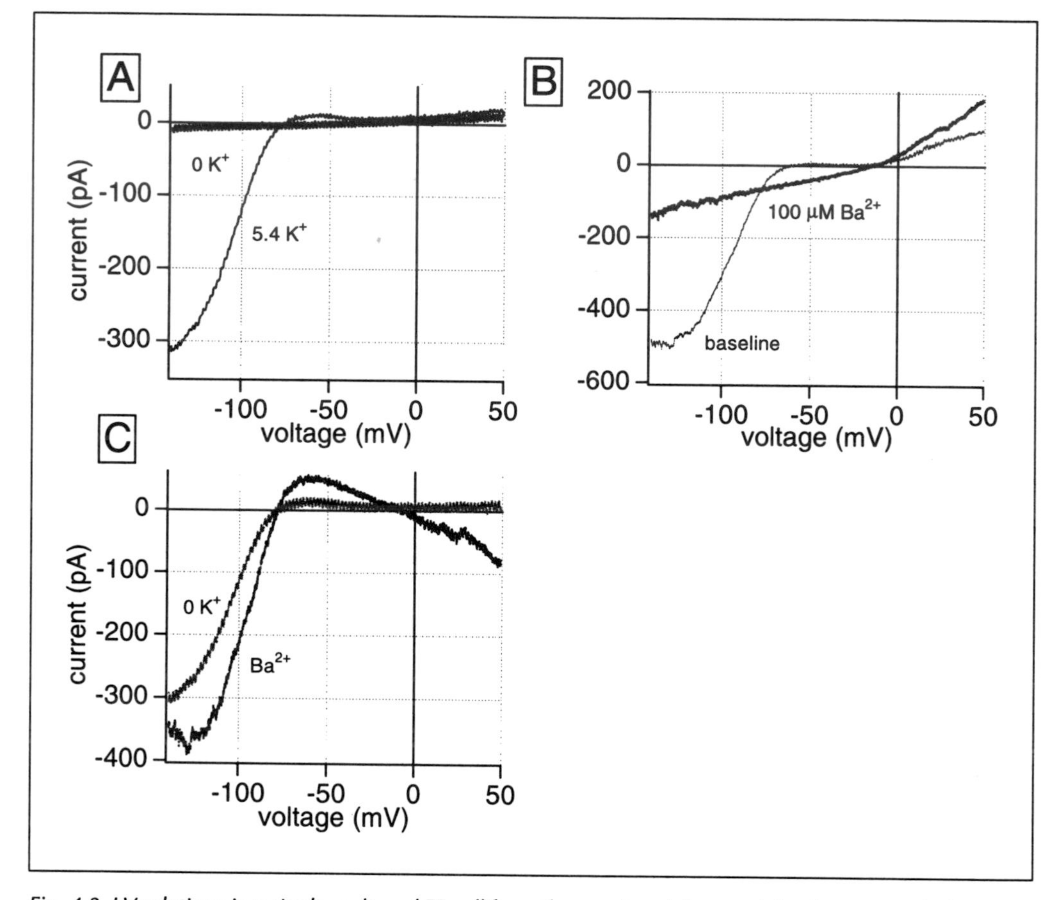

Fig. 4.3. I-V relations in a single, cultured EE cell from the porcine right ventricle after removal of K^+ from the bathing solution (A) and in the presence of 100 µmol/l $BaCl_2$ (B). Ramp clamps, bathing and pipette solutions were as in Figure 4.2, except for $[Cl^-]$ in the pipette (22 mmol/l instead of 42 mmol/l, replacement with aspartate$^-$). Removal of extracellular K^+ depolarized the zero-current potential from -76 to -39 mV, abolished the I_{Ki}-currents, but did not induce or affect outwardly directed currents. Ba^{2+} caused a depolarization of the zero-current potential from -60 to -16 mV and blocked I_{Ki}-currents in a reversible way (wash-out of Ba^{2+}). There was no blocking effect on outwardly directed currents, indicating that these currents are not related to I_{Ki}. The currents, sensitive to removal of K^+ or to the presence of Ba^{2+}, were obtained by point-by-point subtraction of the currents in the absence of K^+ (panel A) or in the presence of Ba^{2+} (panel B) from their respective baseline currents. They are plotted in one figure in panel C. These Ba^{2+}- and K^+-sensitive currents had a similar voltage-dependency and reversed sign at -79 mV. This value is close to E_K, indicating that K^+-currents were affected. Cell capacitances in A and B were respectively 36.2 and 24.2 pF, while the series resistances were 45.9 MΩ (39% compensation) and 24.2 MΩ (38% compensation).

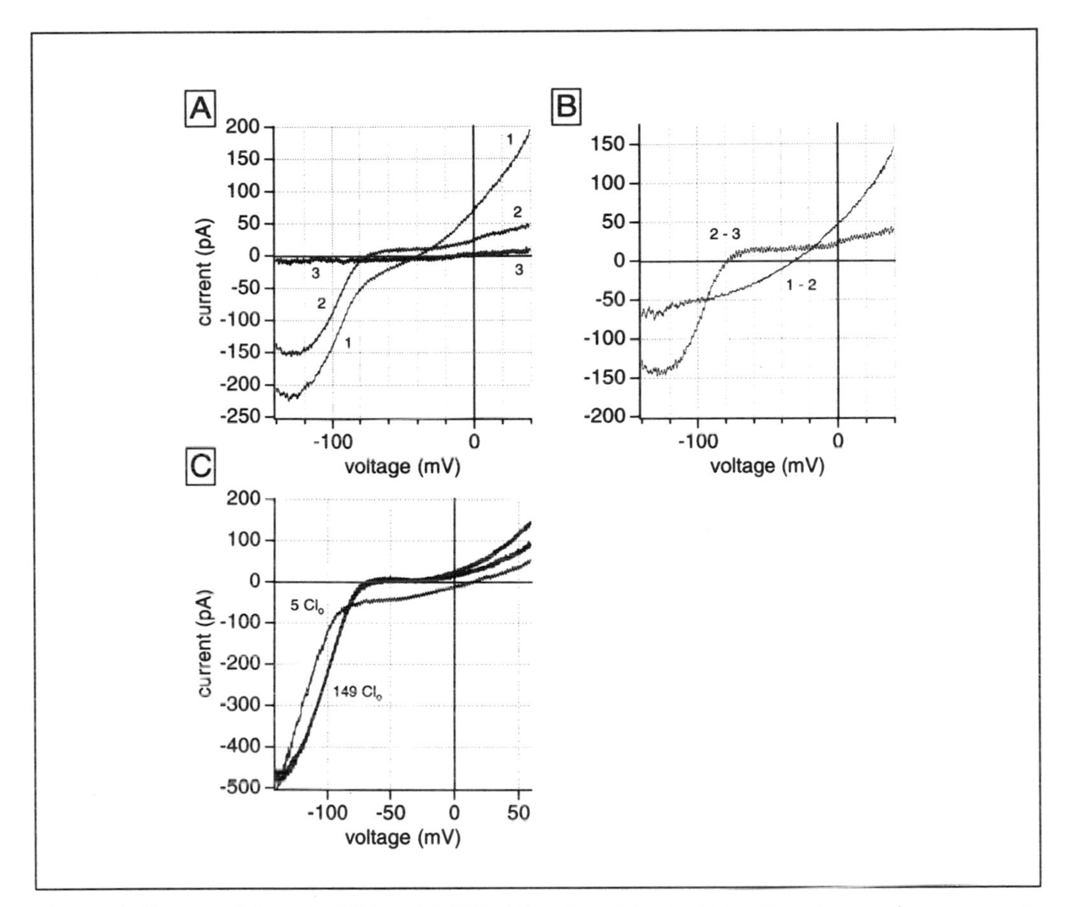

Fig. 4.4. Influence of changing $[Cl^-]_i$ (A, B), $[Cl^-]_o$ (C) and applying $BaCl_2$ (A, B) on the membrane currents (I-V relations) of a single, cultured EE cell from the porcine right ventricle. Clamp protocol, bathing and pipette solutions were as in Figure 4.2. Cellular $[Cl^-]_i$ was changed by whole-cell perfusion of the cell with 22 mmol/l $[Cl^-]$ (trace 1), 5 mmol/l $[Cl^-]$ (trace 2) and 22 mmol/l $[Cl^-]$ plus 10 mmol/l $[Ba^{2+}]$ (trace 3) (A). Upon reduction of $[Cl^-]_i$ from 22 to 5 mmol/l, the zero-current potential hyperpolarized from -40 to -76 mV and especially the outwardly directed currents at positive potentials decreased. If these outwardly directed currents in trace 1 corresponded to inward Cl^--currents, one should expect an increase in outwardly directed currents upon reduction of $[Cl^-]_i$, because of an increase in the driving force for Cl^--ions at positive potentials (E_{Cl} is -97 mVat 5 mmol/l $[Cl^-]_i$; -51 mV at 22 mmol/l $[Cl^-]_i$). In our opinion, the decrease in the outward currents is due to the dilution of some endothelial factor (ATP?) by the pipette perfusion. The subsequent addition of intracellular Ba^{2+} blocked I_{Ki}-currents and depolarized the zero-current potential to -15 mV. By subtraction of trace 2 from trace 1, the current sensitive to the change in $[Cl^-]_i$ was obtained (B). It was an outwardly rectifying current, completely different from the Ba^{2+}-sensitive current (trace 2 - 3) (B), which resembled the I_{Ki}-current as in Figure 4.3C. The reduction in the extracellular $[Cl^-]$ (extracellular Cl^- replaced by gluconate$^-$) from the normal value of 149 mmol/l to 5 mmol/l caused a large depolarization of the zero-current potential as expected for the change in E_{Cl} from -51 mV to +40 mV ($[Cl^-]_i$ was 22 mmol/l). The depolarization was reversed by returning to normal Cl^--concentrations. The reduction of $[Cl^-]_o$ always increased I_{Ki}-currents. This can be explained by the decrease in extracellular Ca^{2+} (bound by gluconate$^-$). Ca^{2+}-removal also increased I_{Ki}-currents. The effects of low $[Cl^-]_o$-solutions on I_{Ki} were reversed upon increase in $[Ca^{2+}]_o$ (not shown). Cell capacitance was 44.8 pF in A and 38.1 pF in C. The series resistance in A was 10.7 MΩ, in C 5.6 MΩ (both 50% compensation).

back to 22 mmol/l, but in the presence of 10 mmol/l Ba^{2+}. In basal conditions, the I-V relation of the cell showed a large inwardly rectifying current (I_{Ki}) and an outwardly rectifying current, which was larger than 2 pA/pF at +60 mV. A reduction in $[Cl^-]_i$ hyperpolarized the cell: a Cl^--dependent outwardly rectifying current (trace 1 - 2 in Fig. 4.4B) was reduced; and the zero-current potential shifted from -40 to -76 mV, following the shift in the equilibrium potential for Cl^--ions from -49 to -95 mV. Subsequent addition of Ba^{2+} abolished I_{Ki} and depolarized the cell to -15 mV. In another cell, I-V relations were measured before, during and after reduction in the extracellular Cl^--concentration (Fig. 4.4C). In this cell, reduction of $[Cl^-]_o$ from 149 to 5 mmol/l (replacement with gluconate), caused a depolarization of the zero-current potential from -66 mV to +18 mV. Also after reduction in $[Cl^-]_o$, the zero-current potential followed the shift in E_{Cl}. Results of these experiments indicate that some cells, besides I_{Ki}, also display an outwardly rectifying Cl^--sensitive current. It is possible that this outwardly rectifying current contributes together with I_{Ki} to the resting potential of EE cells in basal conditions. As a consequence, the zero-current potential of EE cells in our conditions could be determined by a fine balance between I_{Ki}- and Cl^--currents. When the cells were divided into two groups: one group in which outwardly directed currents at +60 mV were smaller than 2 pA/pF and the other group in which these currents were larger than 2 pA/pF, the following observations were made. The mean zero-current potential after correction for seal leakage currents in the first group (I_{out} at +60 mV was 0.80 ± 0.49 pA/pF) was -66.5 ± 8.5 mV (SD, n = 35) and closer to the E_K of -85 mV. In the second group (I_{out} at +60 mV was 6.69 ± 3.56 pA/pF), the mean zero-current potential was -46.5 ± 21.0 mV (SD, n = 17) and closer to the E_{Cl} of -34 mV.

There seems to be a complex regulation of this outwardly rectifying Cl^--current in nonstimulated EE cells. In some cells, the current was not observed; in other cells it disappeared or appeared and increased during cell perfusion. Reasons for these discrepancies in observations might be: (1) some pipette factor; (2) some cellular factor diluted by cell perfusion or (3) different properties of the cells. To our knowledge, there are two reports in vascular endothelial cells (bovine pulmonary artery[42] and human umbilical vein[15]) describing similar outwardly rectifying currents at positive potentials. In the former report, the outwardly rectifying current was not related to I_{Ki} and was not further studied. In the second report, the current was only observed in cells patched with the conventional whole-cell recording technique, but not in cells with permeabilized patches.

In conclusion, the resting membrane conductance of EE cells is mainly determined by K^+-ions. In some experimental conditions, also Cl^--ions contribute in setting the resting potential, indicating that besides K^+-ions also Cl^--ions may play a role in regulating the Ca^{2+}-homeostasis of EE cells.

1.2. Electrophysiology of Stimulated Endocardial Endothelial Cells

Vascular endothelial ion channels, pumps and transporters play an important role in the Ca^{2+}-homeostasis of stimulated endothelial cells. Thereby, they are involved in the release of endothelial-derived factors with an influence on contractility of smooth muscle cells or cardiomyocytes (Fig. 4.5).[1] In vascular endothelial cells, a number of different types of ion channels activated by physical stimuli (membrane stretch, shear stress and cell swelling) or humoral stimuli (bradykinin, histamine, ATP, substance P, acetylcholine, thrombin, endothelin-1 and NO) have been studied. Tables 4.2 and 4.3 give an overview of ion channels in vascular endothelial cells stimulated by these stimuli. In EE cells, stimulus-secretion coupling has not been as extensively studied. How do physical or humoral stimulation of EE cells influence the electrophysiology of these cells?

a. Physical stimulation of endocardial endothelial cells

In cultured EE cells from the porcine right ventricle, membrane stretch, simulated by cell swelling or application of pressure to the patch pipette, activated an outwardly rectifying Cl^--current and concomitant depolarization of the zero-current potential to potentials around the equilibrium potential for Cl^--ions (Fig. 4.6). The outwardly rectifying volume-activated current was not dependent on $[Na^+]_o$ or $[K^+]_o$, but was influenced by changes in $[Cl^-]_o$. The current could be reversed upon going to isosmotic conditions (Fig. 4.6) and could be blocked by DIDS (4,4'-diisothiocyanostilbene-2,2 disulfonic acid, 100 to 300 µmol/l) and flufenamic acid (50 to 100 µmol/l). Furthermore, it was demonstrated that the

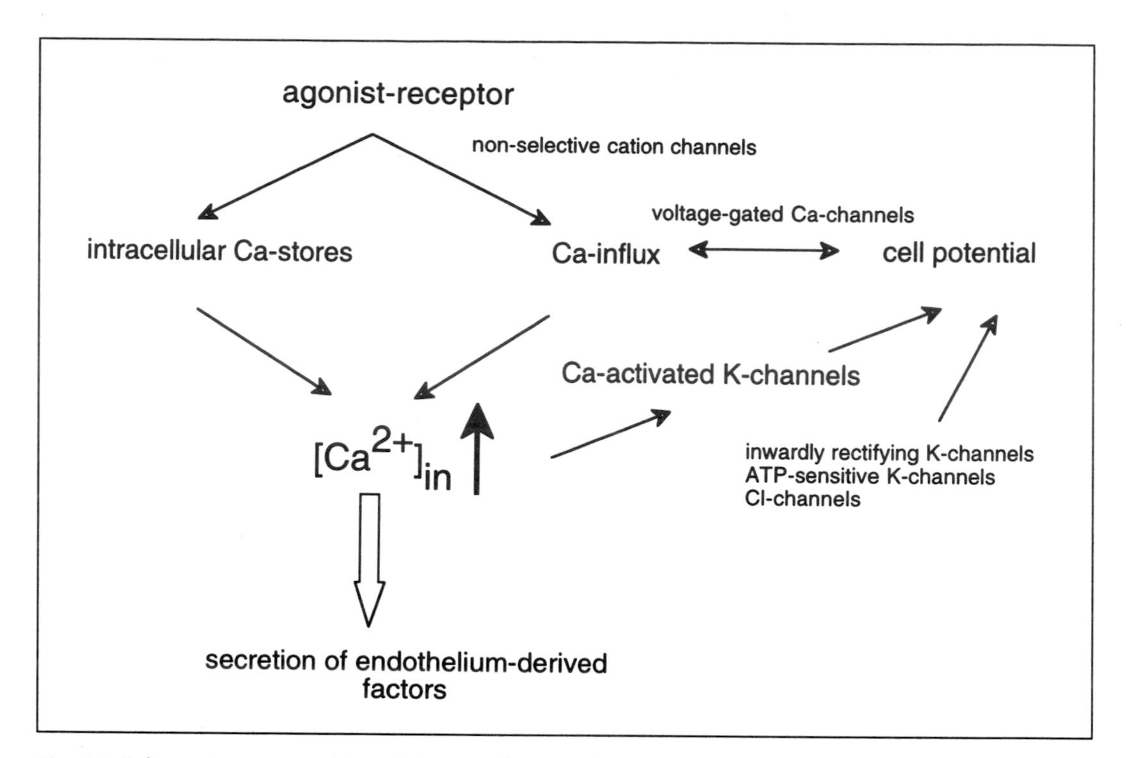

Fig. 4.5. Schematic representation of the contribution of ionic channels to the electrical and Ca^{2+} response of vascular endothelial cells at rest and upon stimulation by vasoactive substances. As a consequence of the agonist-receptor interaction, the intracellular Ca^{2+}-concentration ($[Ca^{2+}]_{in}$) increases. The influx of Ca^{2+}-ions from the extracellular space can occur through nonselective cation channels (directly related to the agonist-receptor interaction) or voltage-gated Ca^{2+}-channels and is modulated by the cell's membrane potential. The cell potential is regulated by inwardly rectifying K^+-channels, Cl^--channels and ATP-sensitive K^+-channels. Because of the increase of $[Ca^{2+}]_{in}$, Ca^{2+}-dependent K^+-channels are activated and the cell hyperpolarizes, leading to a further influx of Ca^{2+}-ions. Modified after Adams.[1]

Table 4.2. Ion channels in vascular endothelial cells activated by physical stimuli

Channel	Agonist	References
I_{cation}	stretch	Lansman et al[20]
		Popp et al[35]
	shear stress	Schwarz et al[40]
	cell swelling	Popp et al[35]
I_K	cell swelling	Perry and O'Neill[34]
	shear stress	Olesen et al[31]
I_{Cl}	cell swelling	Nilius et al[29]

I_{cation}, I_K and I_{Cl} are membrane currents of non-specific cations, K- and Cl-ions. Membrane stretch was caused by applying pressure or suction to the patch-pipette, shear stress by the flow of bathing solution over the cell and cell swelling by application of a hypotonic bathing solution.

Table 4.3. Ion channels in vascular endothelial cells activated by humoral stimuli

Receptor-Operated Channel	Agonist	References
$I_{K(Ca)}$	bradykinin	Colden-Stanfield et al[9,10] Cannel and Sage[7] Mehrke and Daut[26] Sauvé et al[39] Thuringer et al[44] Vaca et al[45] Rusko et al[36]
	ATP	Sauve et al[38] Carter and Ogden[8] Rusko et al[36]
	acetylcholine	Sakai[37] Rusko et al[36]
	histamine	Groschner et al[15]
	substance P	Sharma and Davies[41]
I_{cation}	bradykinin	Cannel and Sage[7] Mendelowitz et al[27]
	endothelin-1	Zhang et al[48]
	thrombin	Johns et al[19]
	histamine	Nilius[28] Graier et al[14] Yamamoto et al[46] Groschner et al[15]
I_K	acetylcholine	Olesen et al[32]
I_{ha} (Na and K)	NO	Janigro et al[18]
I_{Cl}	histamine	Groschner et al[15]

I_{cation}, I_K and I_{Cl} are the currents of non-specific cations, K- and Cl-ions, $I_{K(Ca)}$ is a Ca-activated K-current and I_{ha} a hyperpolarization-activated current of Na- and K-ions.

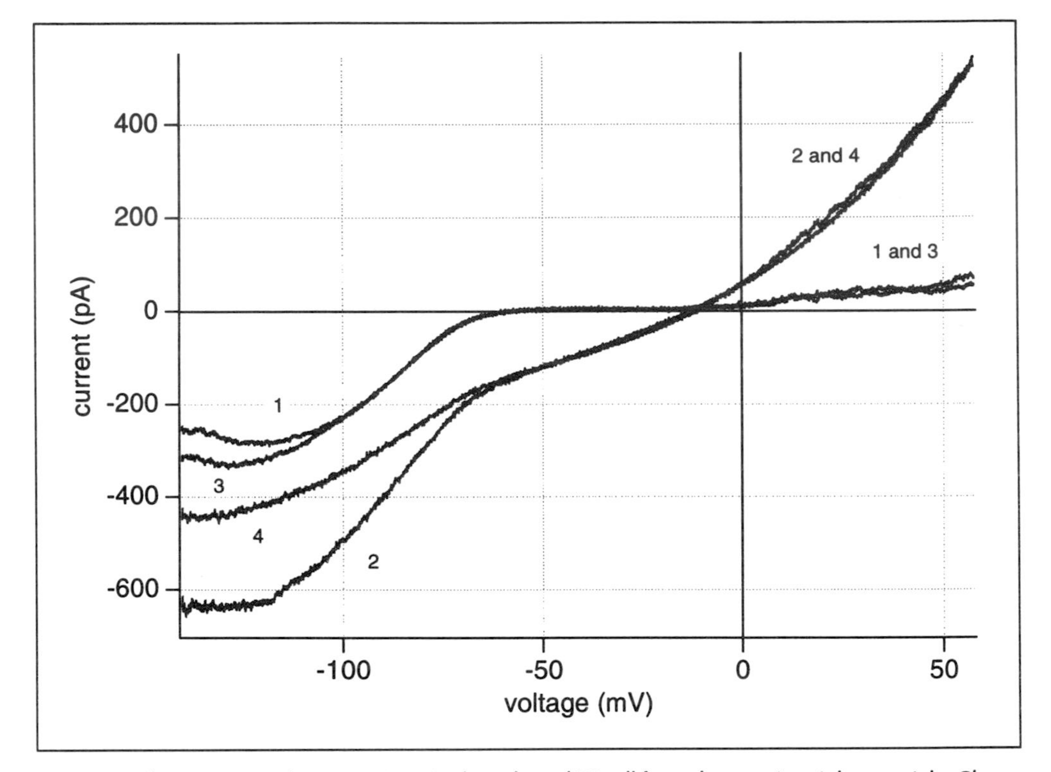

Fig. 4.6. Volume-activated currents in a single, cultured EE cell from the porcine right ventricle. Clamp protocols and bathing solution as in Figure 4.2. Membrane currents after ramp clamps were measured in different conditions: (1) immediately after gaining access to the cell interior with an ATP- (5 mmol/l) containing hypertonic pipette solution (similar solution as in Figure 4.2 but 50 mmol/l K^+ instead of 30 mmol/l): trace 1; (2) 5 minutes after induction of cell swelling with this hypertonic pipette solution: trace 2; (3) 10 to 15 minutes after adding 50 mmol/l sucrose to the bathing solution: trace 3; (4) after adding 40 µmol/l verapamil to the hypotonic bathing solution: trace 4. Cell swelling induced by the perfusion of the cell with the hypertonic pipette solution, induced an increase in inwardly and outwardly directed currents, which stabilized after 5 to 6 minutes. Upon going from condition 1 to condition 2, the cell depolarized from -68 to -13 mV. The currents, induced by hypertonic cell perfusion, were outwardly rectifying and reversed sign at -10 mV, close to the E_{Cl} of -22 mV in this experiment. The induction of this volume-activated outwardly rectifying Cl^--current could be reversed by adding sucrose to the bathing solution to obtain equimolar conditions again (trace 2 to 3) and the cell repolarized to -68 mV. Addition of verapamil, known to block P-glycoprotein related Cl-currents, did not affect the outwardly rectifying Cl-current, ruling out the involvement of MDR1-gene encoded ATP-dependent ABC-transporters (trace 4). Similar phenomena were observed in cells, patched with an ATP-containing isotonic pipette solution but bathed in a hypotonic bathing solution (not shown). Cell capacity was 36.5 pF and the series resistance was 13.4 MΩ (49% compensation). Reprinted Fransen et al, Am J Physiol 1995; 268:H2036-47 with permission from The American Physiological Society.

volume-activated Cl^--current was dependent on the presence of ATP in the pipette solution,[12,13] but that it was not related to the P-glycoprotein-related Cl-channels, as verapamil (40 µmol/l) did not inhibit or prevent the activation of the current (Fig. 4.6). A similar swelling-induced outwardly rectifying current has been described in vascular endothelial cells, where it seemed to be related to MDR (multiple drug resistance)1-gene encoded ATP-dependent ABC-transporters.[29]

There was a remarkable similarity between the swelling-activated Cl^--current and the small outwardly rectifying Cl^--current, observed in 30% of nonstimulated EE cells. The amplitude of the swelling-induced Cl^--current was, however, always much larger than the amplitude of the outwardly rectifying Cl^--current in basal conditions. The dependency of the volume-activated Cl^--current on intracellular ATP might explain why the smaller outwardly rectifying Cl^--current in basal conditions (30 % of the EE cells) was not always observed or why it sometimes increased or decreased during cell perfusion. Endothelial cells produce and release ATP[47] and it is possible that the intracellular ATP concentration in cultured porcine EE cells shows large variability. With respect to this, it would be interesting to investigate the dependency of the outwardly rectifying current in nonstimulated EE cells on the intracellular (pipette) concentration of ATP.

The role of the swelling-activated outwardly rectifying Cl^--current in EE cells is still speculative. It might be involved in the complex processes of cell volume regulation (and regulation of transendothelial permeability). If the Cl^--currents are stretch-dependent or shear stress-dependent, they are expected to influence the membrane potential and to play an important role in the cellular Ca^{2+}-homeostasis and release of endothelium-derived factors. It is even possible that in this situation EE cells, instead of being electrically silent, display some kind of action potentials: depolarizing to E_{Cl} and repolarizing to E_K during each cardiac cycle.

In the EE of the porcine right atrium, the presence of a stretch-activated Ca^{2+}-permeable cation channel was also demonstrated.[16] The activation of this channel simultaneously opened a Ca^{2+}-dependent K^+-channel and, therefore, the influx of Ca^{2+}-ions through the stretch-activated channel was thought to be sufficient to induce an intracellular Ca^{2+}-signal and to activate Ca^{2+}-dependent K^+-channels.

In conclusion, endocardial endothelial cells respond to cell swelling and membrane stretch by the activation of an outwardly rectifying Cl^--current, which was dependent on the presence of intracellular ATP, and the activation of a Ca^{2+}-permeable cation channel. The physiological role of the swelling- and membrane stretch-activated currents is still under investigation.

b. Humoral stimulation of endocardial endothelial cells

In cultured EE cells from bovine atrial valves, addition of bradykinin resulted in the typical biphasic increase in $[Ca^{2+}]_i$, as observed in vascular endothelial cells (Fig. 4.1).[3,21,22] In K^+-free bathing solutions, the addition of bradykinin caused oscillations in V_m and synchronized oscillations in $[Ca^{2+}]_i$. These oscillations were dependent on the influx of Ca^{2+}-ions from the extracellular space. It was suggested that the changes in V_m were the consequence of oscillatory changes in a membrane conductance (possibly nonselective cation channels) that also allows the influx of Ca^{2+} into the cells. In view of our results on the background Cl^--conductance in EE cells, it would be interesting to investigate if Cl^--ions could influence the oscillatory behavior of V_m and $[Ca^{2+}]_i$. A similar Ca^{2+}-behavior was also observed in native EE cell monolayers of rabbit cardiac valves with acetylcholine, bradykinin, histamine and ATP, but not with thrombin.[24]

Some humoral substances like ATP, ADP, AMP, histamine and substance P also activate Ca^{2+}-dependent K^+-channels in EE cells from small tissue preparations of guinea pig hearts, similarly as in vascular endothelial cells (Table 4.3).[25] These authors found no effects of bradykinin, acetylcholine, thrombin, atrial natriuretic peptide, vasopressin or serotonin.

In cultured porcine ventricular EE cells, bradykinin (100 nmol/l) substance P (1 µmol/l) and external ATP (100 µmol/l) were effective in about 50% of the cells. When effective, the agonists always caused a transient depolarization of V_m. The time course of this depolarization was similar for the three agonists: within a few seconds, the cells depolarized from negative resting potentials of -50 to -80 mV to potentials of -40 to +40 mV. The time course of the subsequent repolarization was fast (within 5 seconds) for bradykinin and substance P but much slower (minutes) for ATP. The depolarization was due to the activation of a current with a reversal around 0 mV or slightly positive potentials as shown in

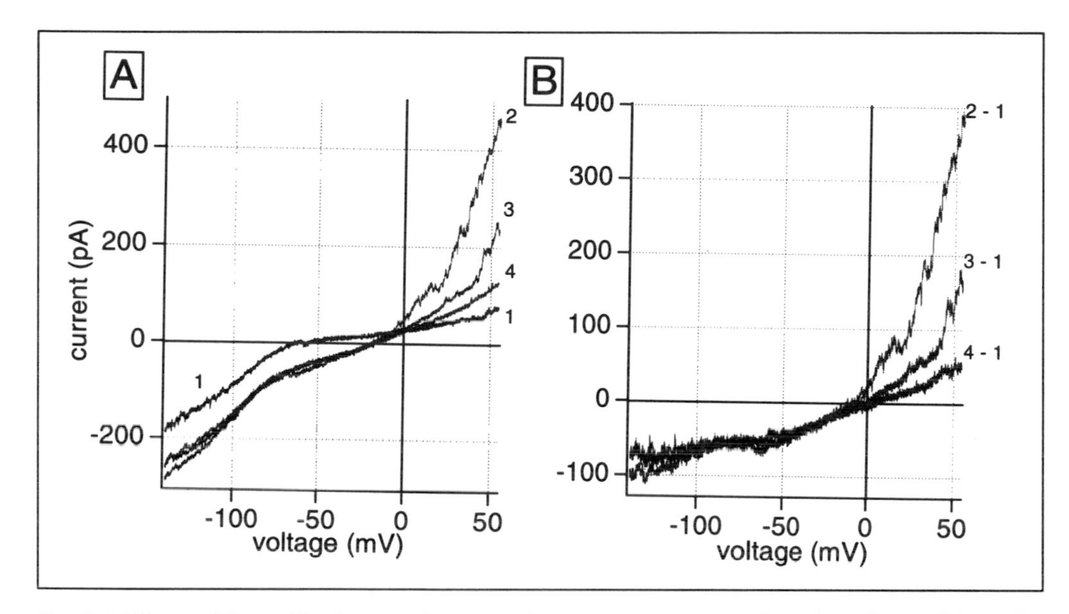

Fig. 4.7. Effects of 1 μmol/l substance P on membrane currents in a single, cultured EE cell from the porcine right ventricle. Ramp clamps and bathing solution as in Figure 4.2. Ramp currents were measured at different time intervals after addition of substance P. Trace 1 is the baseline membrane current in a cell patched with a pipette containing 1 μmol/l $CaCl_2$ and zero EGTA in the pipette (A). Upon addition of substance P, there was a fast increase in inward, but especially outward currents (trace 2, 30 s after trace 1). In the maintained presence of substance P, outwardly directed currents decreased again and almost returned to baseline values after 1 minute (traces 3 and 4). Subtracting trace 1 from the other traces (B) revealed that the initial current activated by substance P was an outwardly rectifying current (trace 2-1), while the current after 1 minute showed ohmic behavior (traces 3-1 and 4-1). The reversal potential of the substance P-sensitive currents was slightly negative (trace 2-1) or slightly positive (traces 3-1 and 4-1). Similar observations were made for bradykinin (100 nmol/l to 1 μmol/l). Cell capacitance was 38.1 pF and series resistance 15.5 MΩ (35% compensation).

Figure 4.7 for substance P. Which type of current is activated by these agonists is still under investigation, but the depolarization of V_m induced by these agonists might be an indication for the activation of nonselective cation channels and/or Cl^--currents, but not K^+-currents. Up to now, we were not able to demonstrate any other K^+-current than I_{Ki} in our preparation. The transient effect of ATP, bradykinin or substance P on V_m seems not to be related to a transient increase in $[Ca^{2+}]_i$, because it was present in Ca^{2+}-buffered (10 mmol/l EGTA) cells as well.

In conclusion, EE cells from different species respond to humoral stimulation. Thereby, the Ca^{2+}-homeostasis and the activation of ion channels is affected. A more detailed study of the influence of agonists on the ion channels of EE cells is necessary to enable a detailed comparison with vascular endothelial cells.

2. BLOOD-HEART BARRIER

In the brain, neuroglia appear to maintain the extracellular concentration of ions, amino acids and other neurotransmitters during firing of the neurons within the narrow limits optimal for the signaling functions of the neurons (short-term regulation). Long-term regulation of the brain extracellular fluid lies in the secretory cells of the choroid plexus and the endothelial cells of the blood-brain barrier.[33] The blood-brain barrier is the best studied endothelial barrier. It is relatively impermeable to ions, many amino acids, small peptides and proteins and has a higher electrical resistance than other endothelia. Regulation of the ionic composition of brain extracellular fluids occurs by a controlled trans-endothelial transport of ions. In essence, there is a transendothelial net Na^+-transport from blood to brain and K^+-transport from brain to blood because of asymmetrical distribution of ion channels, pumps and transporters between luminal and abluminal membrane of the endothelial cells. As a consequence, a specific ionic balance between the blood and the cerebrospinal fluid is created and maintained in order to obtain ideal conditions for brain cell signaling.

As with brain tissue, the myocardium is also a highly excitable tissue, the ionic homeostasis of which is vital not merely to the function of the organ but also to the entire body. In the heart it is, therefore, also essential that the ionic environment of the cardiomyocytes is well controlled over the long term. Alterations in the extracellular concentrations of K^+, Cl^- and Na^+, for example, have been shown to have marked influence on normal rhythmicity and contractility of the heart muscle cells. The unique position of the EE, namely at the interface between blood and cardiomyocytes (not smooth muscle cells) makes it the ideal candidate to regulate the specific ionic environment of the cardiomyocytes. In the following rather speculative part of the chapter, we will discuss several arguments in favor of a blood-heart barrier function of EE cells.

2.1. MORPHO-ELECTRICAL EVIDENCE

EE cells form a thin monolayer of closely apposed cells which are larger and have a more polygonal shape than the endothelial cells of heart blood vessels. Between the cells, there are complex interdigitations, leading to extensive overlap at the junctional edges. This might suggest that the EE has unique permeability properties. Tight junctions are present and always located at the luminal side of the EE. At this luminal side, the glycocalyx is better developed than at the basolateral side below the tight junctions. So, the cells display morphological asymmetry. Gap junctional coupling between EE cells has been demonstrated by connexin-43 labeling and, recently, the spreading of intracellularly applied Lucifer Yellow between neighboring cells was observed (chapter 2). This suggests that the endocardial endothelium may not only be morphologically but also electrically homogeneous. When cells are electrically coupled, it is expected that capacitative currents can be observed upon voltage-clamping small clusters of cells. In cultured porcine EE cells, capacitative currents were determined upon applying short trains of square wave pulses (5 mV) at 0 mV, the potentials at which there is a finite membrane conductance in nonstimulated EE cells (see Fig. 4.2). In these conditions, the capacitative currents give a raw estimate of the membrane capacitance, which is expected to be much larger in electrically coupled cells than in a single cell. In the experiment of Figure 4.8, the membrane capacitance was reduced from 130 to 29 pF upon addition of 1 mmol/l octanol, a substance known for its cell-to-cell uncoupling properties. This might be an indication for the coupling of about four to five cells, assuming that the cells have a similar membrane surface. I-V relations were measured before and after addition of octanol. The currents were about 5 times larger in the cell cluster than in the uncoupled cell. The I-V relations had, nevertheless, a similar form, except that in the cell cluster, there were outwardly directed currents at positive potentials. The outwardly directed currents in the cell cluster might be attributed to Ca^{2+}-dependent currents as the pipette contained 10 mmol/l EGTA, which buffered intracellular Ca^{2+} probably well in the uncoupled cell, but did not diffuse into the cells electrically coupled to this cell. Further investigation of the electrical EE cell-to-cell coupling and the spreading of Lucifer Yellow is planned.

2.2. Evidence from Mechanical Observations

Selective destruction of EE in cat papillary muscles induced characteristic changes in isometric force development (chapter 3). These changes could not be fully reversed by adding several endothelium-derived factors (chapter 5), indicating that the stimulus-secretion-contraction coupling cannot be the only mecha-

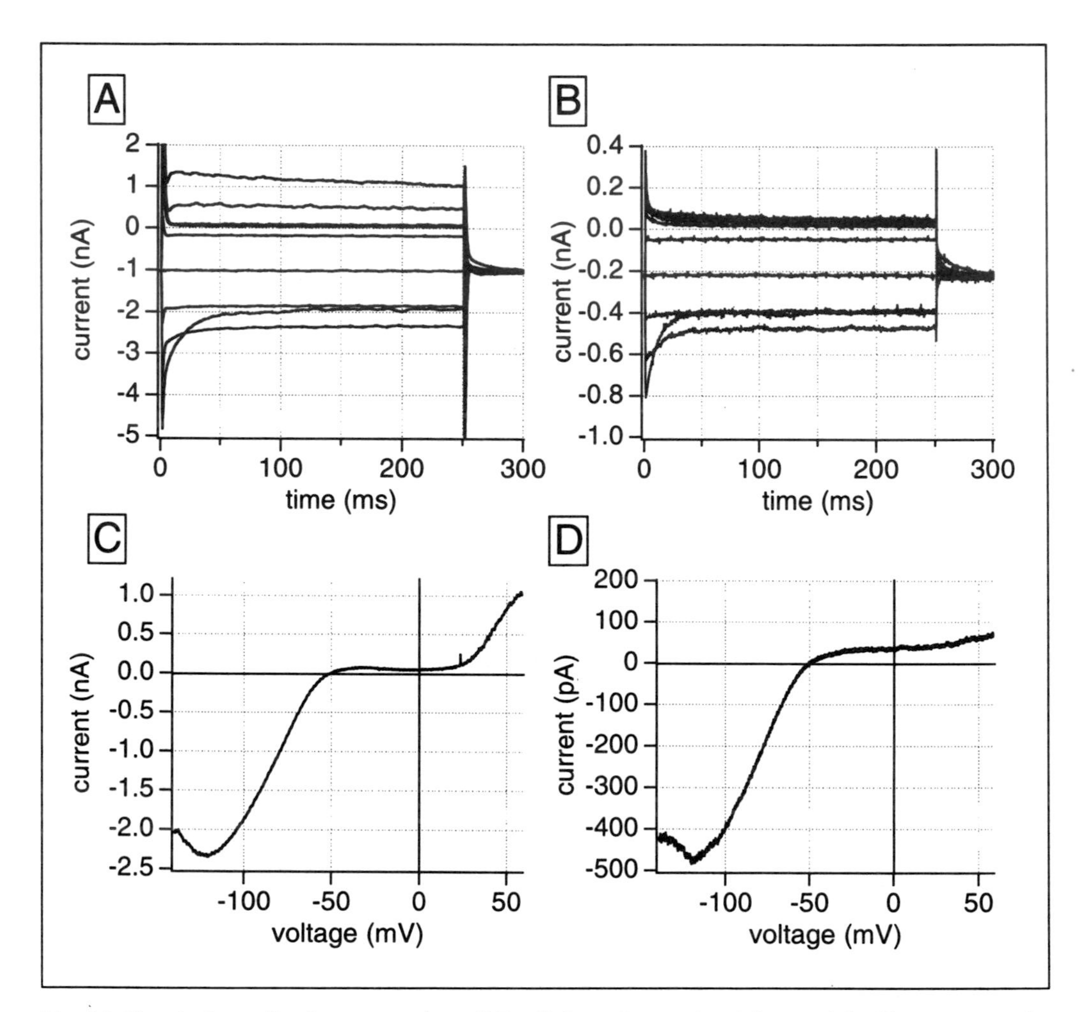

Fig. 4.8. Electrical coupling between cultured EE cells from the porcine right ventricle. Clamp protocols, bathing and pipette solutions were as in Figure 4.2. A: Currents as a function of time for the different clamp steps between -140 and +60 mV were measured in a cell cluster (4 to 6 cells) (panel A) and after uncoupling of the cells by 1 mmol/l octanol (panel B). In a single cell, 1 mmol/l 1-octanol had no effect on membrane currents. In the uncoupled cell, currents were about 5 to 6 times smaller than in the cell cluster (remark the different current scale), but displayed similar time-dependency. The transient capacitive currents during the first 10 ms of the voltage change in the cell cluster were of similar magnitude for the same positive or negative voltage change but in the opposite direction and disappeared after uncoupling the cells with 1-octanol. In panels C and D, ramp clamp currents (and thus I-V relations) in the cluster of cells and after uncoupling of the cells are shown. I-V relations in the cluster and the single cell had a similar form with a reversal potential at -50 mV. In the cluster, an outwardly rectifying current was observed, which was not present in the single cell. This current might be related to changes in $[Ca^{2+}]_i$ in the cells coupled to the patched cell, because only in this cell, $[Ca^{2+}]_i$ was buffered with 10 mmol/l EGTA. Cell capacitance was 130 pF (cell cluster) or 28 pF (single cell) and the series resistance was 5.0 MΩ (75% compensation).

nism accounting for the changes in contractility after EE destruction. Furthermore, the effect of EE destruction on contractility was immediate and there is no reason to assume that all diffusable endothelium-derived factors (especially endothelin) are immediately removed from the extracellular fluids surrounding the cardiomyocytes upon destruction of the EE. On the other hand, removal of a barrier, regulating the ionic milieu of the myocytes, would have an immediate effect on myocardial contractility. With respect to this point, preliminary experiments, in which the influence of EE destruction on isometric force development in cat papillary muscles was studied at different concentrations of K^+- and Cl^--ions in the bathing solution, indicated that the EE might indeed protect the myocardium against high K^+- and low Cl^-- concentrations.

2.3. Evidence from Electrophysiological Data

The strongest evidence for a blood-heart barrier function of the EE comes from the observation that the luminal side of the membrane of EE cells may have a population of ion channels different from the abluminal side. In the luminal membrane of intact EE cells of the bull-frog heart, Ito et al[17] frequently observed non-selective cation channels (an outwardly rectifying and a large conductance ohmic channel). They suggested that K^+-channels should be predominant in the abluminal membrane. If one assumes that the luminal membrane displays cationic channels, while the abluminal membrane expresses a Na^+/K^+ ATPase as in blood-brain endothelial cells, there might indeed be a transendothelial transport of ions, regulating the ionic composition of the extracellular fluid in the heart for optimal electrical activity of the cardiomyocytes. Further investigation is required, especially on the presence of channels, pumps and transporters in the abluminal membrane, to unequivocally demonstrate the presence of a blood-heart barrier.

In conclusion, the electrical coupling between EE cells in clusters and the morphological and electrophysiological asymmetry between luminal and abluminal side of the EE cell membrane highly suggest a possible barrier function of the EE. Direct experimental evidence, demonstrating the ability of EE to regulate the ionic environment of cardiomyocytes, is required to unequivocally show the presence of a functional blood-heart barrier signal transduction system in the EE-cardiomyocyte interaction.

3. GENERAL CONCLUSION

Physicochemical properties of the endocardial endothelium play a role in modulating cardiac performance (Fig. 4.9). In stimulus-secretion-contraction coupling, endocardial endothelial cells respond to agonists by an increase in $[Ca^{2+}]_i$. This response is dependent on the membrane potential of EE-cells, which is modulated by external K^+ and probably also Cl^-. Therefore, both ions may play

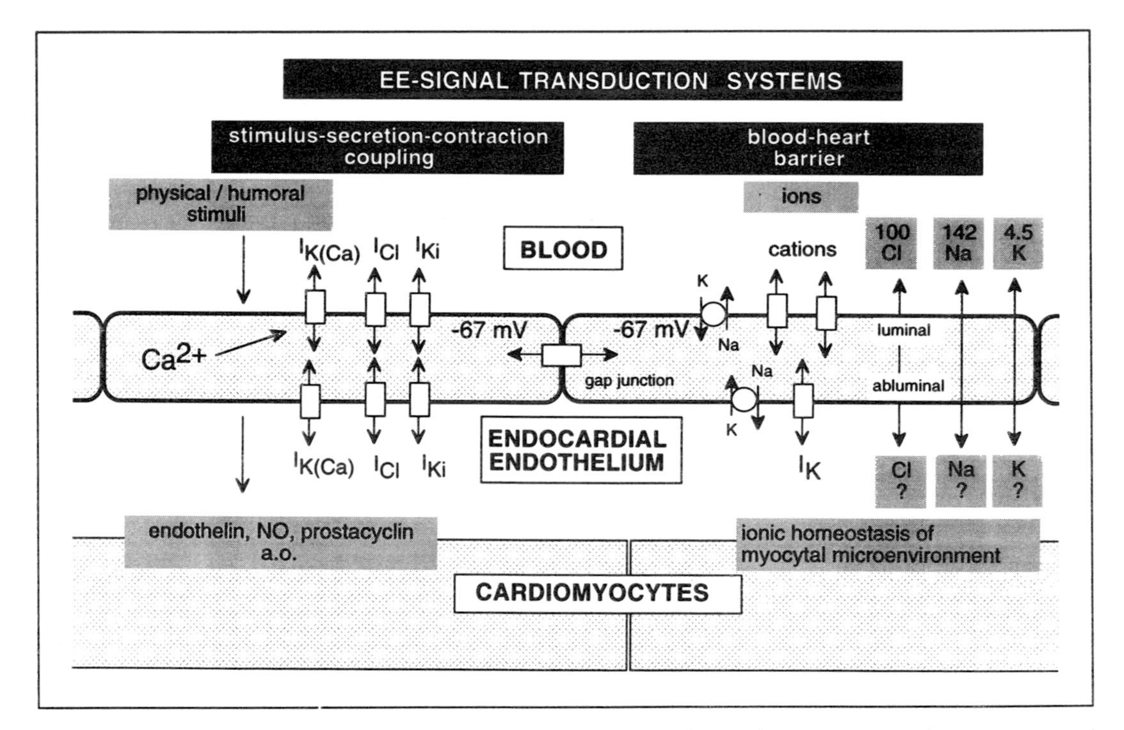

Fig. 4.9. Contribution of ion channels and pumps in the signal transduction systems between EE and cardiomyocytes. EE cells have been shown to respond to physical and humoral stimuli with an increase in $[Ca^{2+}]_{in}$ and activation of different currents.[12,13,16,21,22,23] The Ca^{2+}-response of the cell is further modulated by the activation of Ca^{2+}-dependent K^+-channels ($I_{K(Ca)}$)[16] and by the membrane potential[21,22,23] of the cell in general. The membrane potential is determined by K^+- and Cl^--ions (I_{Ki} and I_{Cl}) and the Na^+/K^+ ATPase.[13,22] These and other electrophysiological events in EE cells are involved in the release of inotropic mediators like endothelin, NO, prostacyclin, and others. Cell-to-cell coupling via gap junctions has been demonstrated (see also chapter 1) and might be important in both signaling systems. Ion channels, pumps and transporters are also expected to be involved in the possible barrier function of the endocardial endothelium. An asymmetrical distribution of ion channels with nonselective cation channels at the luminal, blood-facing side of the membrane and K^+-channels at the abluminal, myocytes-facing side of the membrane has been demonstrated[17] and can lead to a specific transendothelial transport of ions from blood (plasma ion concentrations: 100 mmol/l Cl^-, 142 mmol/l Na^+ and 4.5 mmol/l K^+) to myocytes (ion concentrations in the extracellular space are assumed to be similar as in blood plasma?) and vice versa. Thereby, the EE could represent a long-term regulator of the ionic composition of the myocytal microenvironment similarly as endothelial cells from the blood-brain barrier are long-term regulators of the composition of the cerebrospinal fluids.

an important role in modulating the release of endothelium-derived substances. The EE may also establish some kind of blood-heart barrier, which regulates the ionic composition of the microenvironment of the cardiomyocytes. It will be interesting to compare and evaluate the relative importance of both signal transducing mechanisms in physiological and pathophysiological conditions.

References

1. Adams Dj. Ionic channels in vascular endothelial cells. Trends Cardiovasc. Med 1994; 4:18-26.
2. Adams DJ, Barakeh J, Laskey R et al. Ion channels and regulation of intracellular calcium in vascular endothelial cells. FASEB J 1989; 3: 2389-2400.
3. Adams DJ, Rusko J, Van Slooten G. Calcium signalling in vascular endothelial cells: Ca^{2+} entry and release. In: Weir et al, eds. Ion Flux in Pulmonary Vascular Control. New York: Plenum Press, 1993: 259-275.
4. Busse R, Hecker M, Fleming I. Control of nitric oxide and prostacyclin synthesis in endothelial cells. Arzneim-Forsch / Drug Res 1994; 44: 392-296.
5. Brutsaert DL. The endocardium. Annu Rev Physiol 1989; 51: 263-273.
6. Campbell DL, Strauss HC, Whorton AR. Voltage dependence of bovine pulmonary artery endothelial cell function. J Mol Cell Cardiol 1991; 23: 133-144.
7. Cannell MB, Sage SO. Bradykinin-evoked changes in cytosolic calcium and membrane currents in cultured bovine pulmonary artery endothelial cells. J Physiol (Lond) 1989; 419: 555-568.
8. Carter TD, Ogden D. Kinetics of intracellular calcium release by inositol 1.4.5-triphosphate and extracellular ATP in porcine cultured aortic endothelial cells. Proc R Soc Lond 1992; 250: 235-241.
9. Colden-Stanfield M, Schilling WP, Possani LD et al. Bradykinin-induced potassium current in cultured bovine aortic endothelial cells. J Membrane Biol 1990; 116: 227-238.
10. Colden-Stanfield M, Schilling WP, Ritchie AK et al. Bradykinin-induced increases in cytosolic calcium and ionic currents in cultured bovine aortic endothelial cells. Circ Res 1987; 61:632-640.
11. Daut J, Mehrke G, Nees S et al. Passive electrical properties and electrogenic sodium transport of cultured guinea-pig coronary endothelial cells. J Physiol (Lond) 1988; 402: 237-254.
12. Fransen P, De Keulenaer G, Mohan P et al. The endocardial endothelium: function and electrophysiology. Heart Vessels 1995; Suppl 9:74-76.
13. Fransen PF, Demolder MJM, Brutsaert DL. Whole-cell membrane

currents in cultured endocardial endothelial cells from the porcine right ventricle. Am J Physiol 1995; 268:H2036-H2047.
14. Graier WF, Groschner K, Schmidt K et al. SK&F 96365 inhibits histamine-induced formation of endothelium-derived relaxing factor in human endothelial cells. Biochem Biophys Res Com 1992; 186: 1539-1545.
15. Groschner K, Graier WF, Kukovetz WR. Histamine induces K^+, Ca^{2+} and Cl^- currents in human vascular endothelial cells. Role of ionic currents in stimulation of nitric oxide biosynthesis. Circ Res 1994; 75: 304-314.
16. Hoyer J, Distler A, Haase W et al. Ca2+ influx through stretch-activated cation channels activates maxi K+ channels in porcine endocardial endothelium. Proc Natl Acad Sci USA 1994; 91: 2367-2371.
17. Ito H, Matsuda H, Noma A. Ion channels in the luminal membrane of endothelial cells of the bull-frog heart. Jap J Physiol 1993; 43: 191-206.
18. Janigro D, West GG, Nguyen T-S et al. Regulation of blood-brain barrier endothelial cells by nitric oxide. Circ Res 1994; 75: 528-538.
19. Johns A, Lategan TW, Lodge NJ et al. Calcium entry through receptor-operated channels in bovine pulmonary endothelial cells. Tissue Cell 1987; 19: 733-745.
20. Lansman JB, Hallam TJ, Rink TJ. Single stretch-activated ion channels in vascular endothelial cells as mechanotransducers? Nature 1987; 235: 811-813.
21. Laskey RE, Adams DJ, Cannell M et al. Calcium entry-dependent oscillations of cytoplasmic calcium concentration in cultured endothelial cell monolayers. Proc Natl Acad Sci USA 1992; 89: 1690-1694.
22. Laskey RE, Adams DJ, Johns A et al. Membrane potential and Na^+-K^+ pump activity modulate resting and bradykinin-stimulated changes in cytosolic free calcium in cultured endothelial cells from bovine atria. J Biol Chem 1990; 265: 2613-2619.
23. Laskey RE, Adams DJ, Johns A et al. Regulation of $[Ca^{2+}]_i$ in endocardial cells by membrane potential. In: Rubanyi GM, Vanhoutte PM, eds. Endothelium-Derived Relaxing Factors, Basel: Karger, 1990: 128-135.
24. Laskey RE, Adams DJ, Van Breemen C. $[Ca^{2+}]_i$ measurements in native endothelial monolayers of rabbit cardiac valves using imaging fluorescence microscopy. Am J Physiol 1994; 266: H2130-H2135.
25. Manabe K, Ito H, Matsuda H et al. Hyperpolarization induced by vasoactive substances in intact guinea-pig endocardial endothelial cells. Heart Vessels 1995; Suppl 9:77-79.
26. Mehrke G, Daut J. The electrical response of cultured guinea-pig coronary endothelial cells to endothelium-dependent vasodilators. J Physiol Lond 1990; 430: 251-272.

27. Mendelowitz D, Bacal K, Kunze D. Bradykinin-activated calcium-influx pathway in bovine aortic endothelial cells. Am J Physiol 1992; 262: H942-H948.
28. Nilius B. Regulation of transmembrane calcium fluxes in endothelium. NIPS 1991; 6: 110-114.
29. Nilius B, Oike M, Zahradnik I et al. Activation of a Cl^- current by hypotonic volume increase in human endothelial cells. J Gen Physiol 1994; 103: 787-805.
30. Northover BJ. The membrane potential of vascular endothelial cells. Adv Microcirc 1980; 9: 135-160.
31. Olesen S-P, Clapham DE, Davies PF. Haemodynamic shear stress activates a K^+ current in vascular endothelial cells. Nature 1988; 331: 168-170.
32. Olesen S-P, Clapham DE, Davies PF. Muscarinic-activated K^+ current in bovine aortic endothelial cells. Circ Res 1988; 62: 1059-1064.
33. Orkand RK, Opava SC. Glial function in the homeostasis of the neuronal micro-environment. NIPS 1994; 9: 265-267
34. Perry PB, O'Neill WC. Swelling-activated K fluxes in vascular endothelial cells: volume regulation via K-Cl cotransport and K channels. Am J Physiol 1993; 265: C763-769.
35. Popp R, Hoyer J, Meyer J et al. Stretch-activated non-selective cation channels in the antiluminal membrane of porcine cerebral capillaries. J Physiol Lond 1992; 454: 435-449.
36. Rusko J, Tanzi F, Van Breemen C et al. Calcium-activated potassium channels in native endothelial cells from rabbit aorta: conductance, Ca^{2+} sensitivity and block. J Physiol Lond 1992; 455: 601-621.
37. Sakai T. Acetylcholine induces Ca-dependent K currents in rabbit endothelial cells. Jap J Pharmacol 1990; 53: 235-246.
38. Sauvé R, Parent L, Simoneau C et al. External ATP triggers a biphasic activation process of calcium-dependent K^+ channel in cultured bovine aortic endothelial cells. Pflügers Arch 1988; 412: 469-481.
39. Sauvé R, Chahine M, Tremblay J et al. Single-channel analysis of the electrical response of bovine aortic endothelial cells to bradykinin stimulation: contribution of a Ca^{2+} dependent K^+ channel. J Hypertension 1990; 8: S193-201.
40. Schwarz G, Droogmans G, Nilius B. Shear stress induced membrane currents and calcium transients in human vascular endothelial cells. Pflügers Arch 1992; 421: 394-396.
41. Sharma NF, Davies MJ. Mechanism of substance P-induced hyperpolarization of porcine coronary artery endothelial cells. Am J Physiol 1994; 266: H156-H164.
42. Silver MR, DeCoursey TE. Intrinsic gating of inward rectifier in bovine pulmonary artery endothelial cells in the presence or absence of Mg^{2+}. J Gen Physiol 1990; 96: 109-133.

43. Takeda K, Schini V, Stoeckel H. Voltage-activated potassium, but not calcium currents in cultured bovine aortic endothelial cells. Pflügers Arch 1987; 410: 385-393.
44. Thuringer D, Diarra A, Sauvé R. Modulation by extracellular pH of bradykinin-evoked activation of Ca^{2+}-activated K^+ channels in endothelial cells. Am J Physiol 1991; 261: H656-H666.
45. Vaca L, Schilling WP, Kunze DL. G-protein-mediated regulation of a Ca^{2+}-dependent K^+ channel in cultured vascular endothelial cells. Pflügers Arch 1992; 422: 66-74.
46. Yamamoto Y, Chen G, Miwa K et al. Permeability and Mg^{2+} blockade of histamine-operated cation channel in endothelial cells of rat intrapulmonary artery. J Physiol Lond 1992; 450: 395-408.
47. Yang S, Cheek DJ, Westfall DP et al. Purinergic axis in cardiac blood vessels. Agonist-mediated release of ATP from cardiac endothelial cells. Circ Res 1994; 74: 401-407.
48. Zhang H, Inazu M, Weir B et al. Endothelin-1 inhibits inward rectifier potassium channels and activates nonspecific cation channels in cultured endothelial cells. Pharmacology 1994; 49: 11-22.

CHAPTER 5

Mechanism of Endocardial Endothelial Modulation by Signal Transduction

Puneet Mohan, Gilles W. De Keulenaer and Stanislas U. Sys

There is now compelling evidence to suggest that the endocardial endothelium (EE) modulates performance of the subjacent myocardium and plays an important role in regulation of cardiac function.[5] EE modulation of myocardial contraction has been confirmed in a number of different species and both in vitro and in vivo conditions.[5,15,24,43,53] The mechanisms of EE modulation of myocardial performance are still under investigation and the possibilities include role of EE as a physico-chemical barrier and/or release of various chemical messengers by the EE (Fig. 5.1). As selective damage of EE leads to abbreviation of cardiac twitch duration and concomitant decrease in peak twitch force and shortening,[5,24,43,53] the presence of an intact EE imparts a positive inotropic effect on the subjacent myocardium. The proposed mechanisms underlying the endocardial endothelial regulation as yet, on their own, do not fully explain this tonic action of EE.

In the early experiments, EE was demonstrated to modulate the inotropic response to various substances and cells. While the inotropic effect of atrial natriuretic peptide was dependent on a morphologically intact EE,[33] the myocardial effects of platelets, ATP, and serotonin were enhanced after morphological damaging

Endocardial Endothelium: Control of Cardiac Performance, edited by Stanislas U. Sys and Dirk L. Brutsaert.

of the EE.[47,50] The positive inotropic effect of the α_1-adrenoceptor agonist, phenylephrine, in low physiological concentrations (10^{-9}-10^{-7}M), also required presence of an intact EE to manifest.[32] The EE modulated the inotropic effects of vasopressin; a negative inotropic effect with abbreviation of twitch in presence of intact EE which contrasted with a positive inotropic effect after selective damaging of EE.[45] On the basis of these experiments, it was hy-

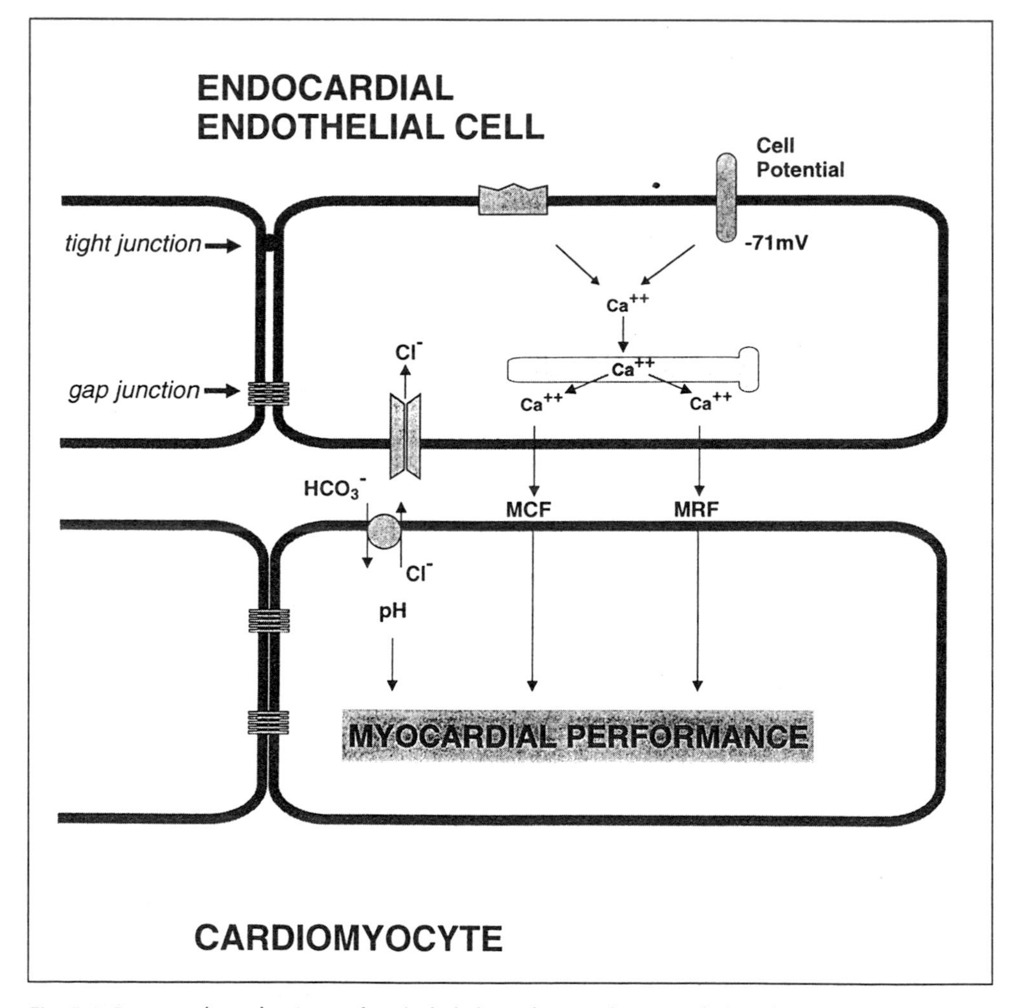

Fig. 5.1. Proposed mechanisms of endothelial regulation of myocardial performance. Presence of long intercellular clefts and numerous gap and tight junctions between cells of EE support the hypothesis of a transendothelial physico-chemical control by EE. EE may also regulate certain ionic currents (e.g. Cl^-) in the cardiomyocyte. Analogous to vascular endothelium, it was hypothesized that EE may release myocardial contraction (MCF) and relaxation (MRF) factor(s) in response to different inotropic stimuli, which would regulate myocardial performance.

pothesized that EE may release myocardial contraction (MCF) and relaxation (MRF) factor(s) in response to different inotropic stimuli[6] (Fig. 5.1). For example, the EE-dependent positive inotropic effect of low concentration of phenylephrine[32] would be due to the release of MCF from the EE, whereas the EE-dependent atrial natriuretic peptide-induced early relaxation and fall in peak twitch tension[33] would result from release of MRF from the EE. The responses to other substances like serotonin,[47] vasopressin,[45] angiotensin[6] and endothelin[25] may be thus explained by alterations in the release and/or balance of MCF and MRF in the EE. Recently, a change in contractility of isolated trabecula by superfusion with coronary venous effluent from an isolated perfused heart was observed which depended on the degree of oxygen saturation of the effluent and the rate of coronary flow.[43] The response was completely eliminated if the endothelial cells in both the perfused heart and the superfused trabecula were damaged and was altered if the EE was damaged. The results, thus, suggested that the effluent contained 'pre-endothelial factors' which signal the endocardial endothelial cells on the trabecula to release 'endothelial factors' which alter the contractility of the trabecula. This would also suggest the existence of a sensory role for the EE for various substances or cells and maybe even for changes in cavitary volume. The presence of several types of receptors in EE, e.g., receptors for atrial natriuretic peptide, supports this hypothesis. Thus the EE can be postulated to play an important role in signal transduction to the subjacent myocyte and thereby regulate its contractile performance.

The vascular endothelium takes part in the regulation of vascular tone by releasing relaxing and contracting factors under basal conditions and when activated by neurotransmitters, hormones, autacoids or physical stimuli. These factors include nitric oxide (NO) which accounts for the biological properties of the endothelium-derived relaxing factor, the potent vasodilator prostacyclin (PGI_2) and the vasoconstrictor endothelin. Similarly, there is now increasing experimental evidence that EE cells release chemical mediators which may determine the myocyte inotropic response. Effluent from superfused cultured porcine endocardial endothelial cells was bioassayed with endothelium-denuded pig coronary artery rings or with EE-denuded ferret papillary muscles.[51,53] The effluent caused vasodilation of the de-endothelized coronary artery. This response could be potentiated in the presence of superoxide dismutase and catalase while it was blocked by hemoglobin and inhibitors of nitric oxide synthase. Thus cultured endocardial cells released an endothelium-derived relaxing factor (EDRF)-like (nitric oxide?) substance. Indeed, EE has now been shown to be capable of releasing *nitric oxide* (NO), *prostaglandins* (prostacyclin and PGE_2) and *endothelin* (Fig. 5.5). In addition, release of a unique 'myofilament desensitizing factor' from cultured endocardial endothelial cells has also been reported.[49]

1. ROLE OF NITRIC OXIDE-CYCLIC GMP

The demonstration of Ca^{2+}-dependent NO synthase in EE cells[46] supports the contribution of NO in EE modulation. Further evidence for a role for NO in EE modulation has come from experiments utilizing substance P, a known stimulant of NO release from vascular endothelium. In isolated ferret papillary muscles with intact EE, substance P induced an abbreviation of twitch contraction similar in pattern to EE removal, and an increase in myocardial cyclic GMP (cGMP).[53] This effect required the presence of intact EE and was inhibited by hemoglobin, an inhibitor of NO activity, thus suggesting that it was mediated by the release of NO from the EE. Continuous basal release of EDRF by EE cells from isolated cardiac valves has also been demonstrated.[21] In the isolated papillary muscle with intact EE a variety of agents that elevate intracellular cGMP either directly like atrial natriuretic peptide[33] or through release of NO like sodium nitroprusside[33,53] had similar effects on myocardial performance as substance P. Utilization of relatively high concentrations of 8-bromo cGMP, an analogue of cGMP, also caused a similar abbreviation of twitch associated with some decrease in contractile performance.[48,52] These early experiments focused on the similarity of the pattern of twitch abbreviation and concomitant decrease in twitch tension observed with selective damage of EE and measures that elevated myocardial cGMP concentration. There was, however, no evidence as to how the EE regulated the myocardial cGMP concentration. Further the inotropic effects of NO-cGMP, as described in the literature, could not fully explain as to how their basal release would account for the unique positive inotropic effect associated with prolongation of the twitch duration, imparted by the presence of intact EE.

Recently, in a preliminary report,[36] we reported for the first time, a novel concentration-dependent positive inotropic effect of NO-releasing nitrovasodilators in muscles with damaged EE which contrasted with their negative inotropic effect in muscles with intact EE (Fig. 5.2 and 5.3) and both these opposite responses were mediated by cGMP. There was a biphasic effect on myocardial performance with administration of 8-bromo cGMP. Whereas addition of low concentrations of 8-bromo cGMP induced a positive inotropic effect, subsequent higher concentrations had a negative inotropic effect (Fig. 5.2 and 5.3). Damaging the EE shifted the exogenous cGMP concentration-tension curve to the right. The concentration-dependent biphasic inotropic response to exogenous cGMP was mimicked by administration of S-nitroso-acetyl-penicillamine, a stable and potent NO-donor, in muscles with intact EE. These results suggest that the inotropic response to NO and cGMP is modulated by the status of the EE. An attractive hypothesis may be formulated from these observations suggesting regulation of intracellular cGMP concentration in the cardiomyocyte by the overlying EE. An intact EE would induce higher cGMP level in the underlying cardiomyocytes due to basal NO release while damaging the EE would lower the cGMP level. The results suggest that in the presence of an intact EE, nitrovasodilators or exogenous nitric oxide cause a further elevation in already high baseline myocardial cGMP levels leading to a negative inotropic response. In case of damaged, as in these experiments, or dysfunctional EE, a similar increase in the myocardial cGMP by exogenous nitric oxide donors from a lower baseline myocardial cGMP level would result in a positive inotropic response. The precise level of exogenous or stimulated cGMP at which the direction of the inotropic response changes from positive to negative, may depend on the basal cGMP levels which in turn may be determined by the state of the overlying EE. In blood vessels, tissue cGMP levels have been frequently reported to be significantly higher in endothelium-intact than in denuded vessels (for references see 17), although this feature has not yet been confirmed in cardiac tissue.[53]

The results demonstrate, for the first time that relatively smaller elevations in cGMP level have a positive inotropic effect and this was further confirmed by a study examining effects of a cGMP analogue in vivo.[23] In acutely instrumented cats, administration of dibutyryl cGMP, a cGMP analogue, in the absence of changes in

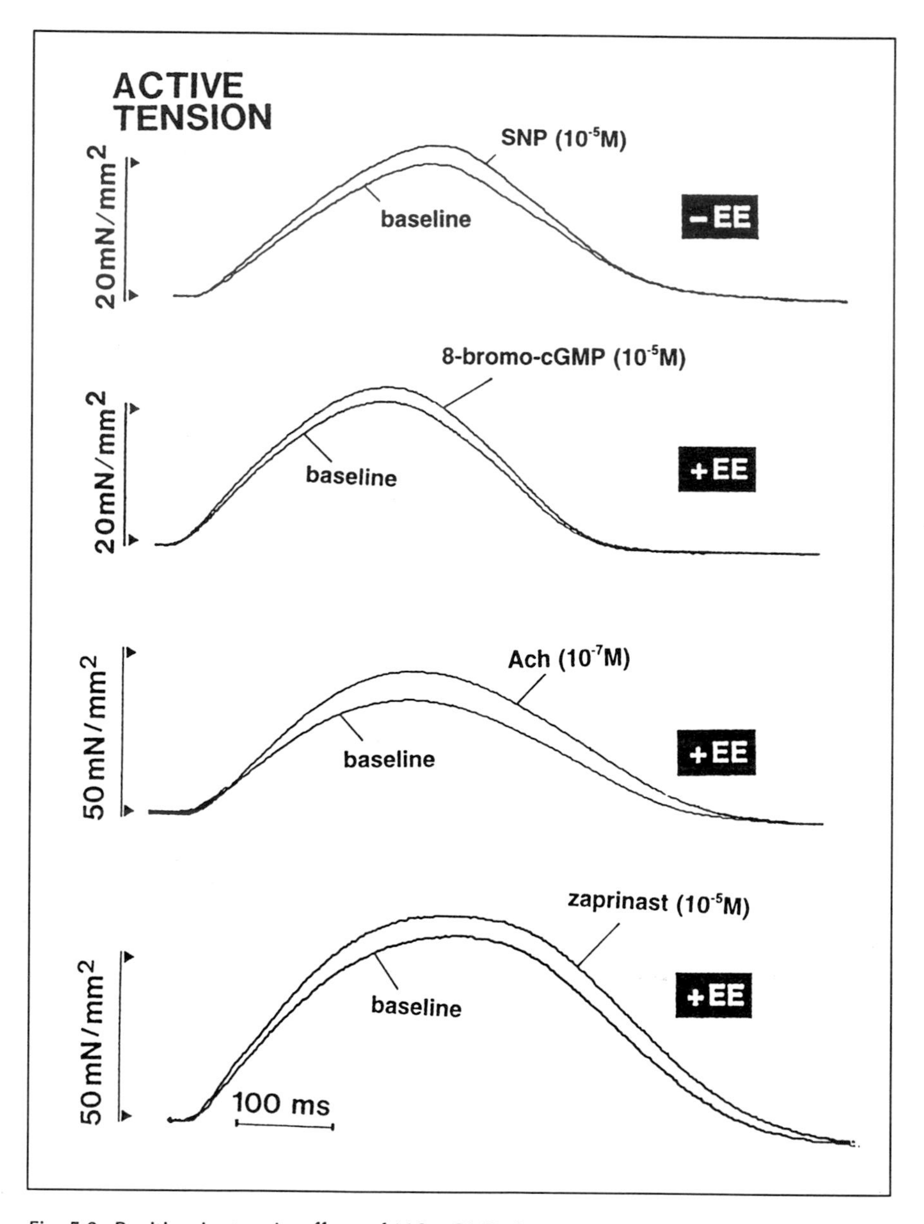

Fig. 5.2. Positive inotropic effect of NO-cGMP. Representative examples of isometric twitches obtained from isolated cat papillary muscle (Krebs-Ringer; Ca^{2+} 1.25mM; 35 °C) showing positive inotropic effect of sodium nitroprusside (SNP; 10^{-5} M), acetylcholine (Ach; 10^{-7} M), 8-bromo-cGMP (10^{-5} M) and zaprinast (10^{-5} M). +EE, intact endocardial endothelium; -EE, damaged EE (exposure to Triton X-100 (0.5%) for 1s).

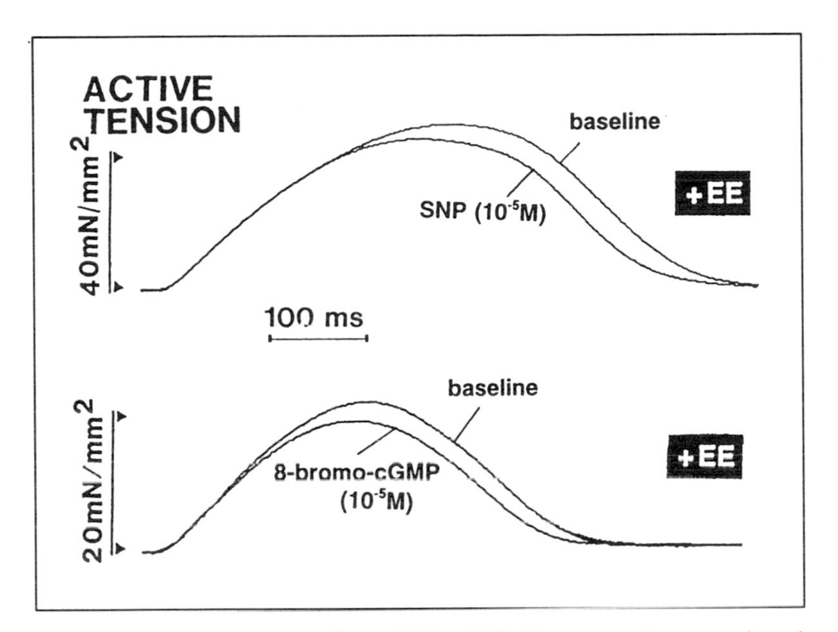

Fig. 5.3. Negative inotropic effect of NO-cGMP. Representative examples of isometric twitches obtained from isolated cat papillary muscle (Krebs-Ringer; Ca^{2+} 1.25mM; 35°C) showing negative inotropic effects of sodium nitroprusside (SNP; 10^{-5} M) and 8-bromo-cGMP (10^{-3} M). +EE, intact endocardial endothelium.

ventricular loading conditions, caused a positive inotropic effect with no effect on parameters of relaxation. In isolated papillary muscles cGMP was also suggested to account for at least a part of the positive inotropic effect induced by acetylcholine[35] (Fig. 5.2). On the basis of these results, we, therefore, suggest that basal release of NO from the EE and the consequent elevation in cGMP in the cardiomyocyte confers a positive inotropic effect. A NO synthase inhibitor-induced cardiac depression in animals without associated change in arterial pressure, heart rate,[19,20] coronary flow and oxygen supply-demand ratio,[19] and regional myocardial tissue perfusion[44] may be explained partly by a significant reduction in myocardial cGMP. Recently two separate groups reported an unexplained NO synthase-dependent myocardial depression.[16,54] Inhibition of NO synthesis worsened myocardial stunning independent of effects on blood flow in conscious dogs.[16] In human volunteers, inhibition of basal NO release with infusion of L-NMMA was associated with a decline in cardiac output and stroke volume.[54] The authors suggested that some basal release of

NO is required to preserve cardiac function in vivo. We have further evaluated the effect of increase in basal intracellular cGMP concentration in myocardium by inhibiting the degradation of cGMP, by utilizing zaprinast (M&B 22948), a selective inhibitor of cGMP phosphodiesterase (PDE V). Zaprinast, between 10^{-9} M and 10^{-5} M, caused a concentration-dependent positive inotropic effect with no change in twitch duration (Fig. 5.2). These results are similar to those obtained with lower concentrations of exogenous cGMP administration both in vitro and in vivo. Damaging the EE shifted the terminal portion of the concentration response curve in a similar manner to the shift observed of the concentration response curve with 8-bromo cGMP. Thus, irrespective of the method, elevation of intracellular cGMP concentration within a physiological range, results in a positive inotropic effect. These results are also in agreement with previous unexplained reports of positive inotropic effect with zaprinast in vivo.

The mechanism(s) as to how cGMP mediates the positive inotropic effect can only be speculated (Fig. 5.5). Based on current knowledge, the response to cGMP could be mediated by a recently described new second messenger, cyclic ADP ribose (cADPR), which stimulates release of Ca^{2+} from intracellular stores through ryanodine receptor in sea urchin eggs.[3,7,12,14] It has also been reported that cADPR increase the open probability of cardiac ryanodine-sensitive Ca^{2+} channels and thus cADPR can trigger the release of Ca^{2+} from the sarcoplasmic reticulum in cardiac cells.[39] In sea urchin eggs, cGMP-induced Ca^{2+} transient was mediated through ryanodine receptors by stimulation of cADP-ribosyl cyclase.[13] Stimulation of guanylate cyclase by NO causes elevated cGMP which stimulates cGMP-protein kinase (PKG). PKG phosphorylates ADP-ribosyl cyclase, enhancing the synthesis of cADPR from NAD^{+} which then stimulates release of calcium from sarcoplasmic reticulum through ryanodine receptor.[7] The overall result would be an increase in cytosolic calcium concentration that could explain the positive inotropic effect of basal NO and cGMP.

Another possible mechanism underlying the inotropic effects of NO and cGMP could involve modulation of cGMP-dependent cAMP phosphodiesterase by cGMP. The evidence for this mechanism comes from electrophysiological studies on isolated cardiomyocytes. In isolated guinea pig ventricular myocytes, a stimulatory effect of relatively low concentrations of cGMP (from 10^{-7} M to 10^{-5} M) on cAMP-elevated L-type Ca^{2+} channel current (I_{Ca}) has been reported[40] leading to an increase in Ca^{2+} availability. Higher concentrations of cGMP, 8-bromo-cGMP, or cGMP-dependent protein kinase (cGMP-PK) had either no effect or reduced I_{Ca}. It was suggested that this stimulation of cAMP-elevated I_{Ca} by low concentrations of cGMP was due to participation of cGMP-inhibitable cAMP-phosphodiesterase, whose presence in the heart has been well documented.[2,56] The higher concentrations of cGMP have been reported to inhibit cAMP-elevated I_{Ca} via cGMP-PK in mammalian myocytes; intracellular perfusion of cGMP-PK fragment caused a similar inhibition of I_{Ca}.[37] In addition to its direct effects on Ca^{2+} channel, cGMP-PK may also decrease the Ca^{2+} sensitivity of the myofilaments through phosphorylation of the inhibitory subunits of troponin.[42] In isolated cardiac myocytes, administration of relatively high concentrations of 8-bromo cGMP (5 x 10^{-5} M) was associated with a negative inotropic effect which was mediated by cGMP-PK-induced decreased Ca^{2+} sensitivity of the myofilament.[52] Recently a NO-mediated biphasic response on stimulated I_{Ca} was reported in frog ventricular myocytes with administration of increasing concentrations of SIN-1.[38] All the responses to SIN-1 were inhibited by methylene blue and by LY83583, another inhibitor of guanylate cyclase. The authors suggested that the stimulatory effect of NO donors on I_{Ca} resulted from an inhibition of the cGMP-inhibitable cAMP-phosphodiesterase while the inhibitory response was due to activation of the cGMP-stimulated cAMP-phosphodiesterase, both linked to the activation of guanylate cyclase. Similar stimulatory and inhibitory effects of SIN-1 on I_{Ca} were also reported by another group but both the effects were said to be mediated by cGMP-PK.[9]

To summarize, EE probably regulates basal intracellular cGMP concentration in cardiomyocytes by basal release of NO. Basal or tonic release of NO from the EE and consequent elevation in cardiomyocyte cGMP concentration, by virtue of its positive inotropic response, will thus be important in preserving cardiac function in physiological states. However, excess cGMP stimulation due to large amount of NO released from nitrovasodilators or from inducible NO-synthase in pathological states would cause a negative inotropic effect and contribute to deterioration of cardiac function.

2. ROLE OF PROSTAGLANDINS

As in vascular endothelium, release of prostacyclin (PGI_2) from the EE may also play an important role in the modulation of myocardial function. Tissue cyclo-oxygenase has been found to be twice as high in the endocardium as compared to the myocardium and located specifically in the endothelial fraction.[4] EE from isolated cardiac valves[21] as well as valvular EE cells in culture[27] produce significant amounts of prostacyclin. Recently, production of both prostacyclin and prostaglandin E_2 from cultured bovine right and left ventricular EE cells was reported with prostacyclin production 10 times greater than prostaglandin E_2.[30] It is also of interest that EE cells have been reported to produce far larger amounts of prostacyclin than vascular endothelial cells.[29]

The role of endogenous prostaglandins, released from the EE in large amounts, has been difficult to explain in the past. Previous studies utilizing exogenous prostaglandins to examine their direct myocardial actions and the underlying mechanisms, mainly cAMP-mediated, were unclear; the responses to prostacyclin and prostaglandin E_2 ranged from increased contractility in isolated atria, to no effect or negative inotropic effect in isolated papillary muscles.

We recently postulated that although endogenous NO and prostaglandins released from the EE may significantly contribute to the endothelial modulation of myocardial function, their mutual interaction may be at least as important. We examined the interactions of endogenous NO and prostaglandins released from the EE and how this interaction affects the myocardial performance in isolated papillary muscles.[34] Papillary muscles were divided into two groups. In the first group substance P, which causes release of NO from vascular endothelium, was utilized to stimulate release of NO from the EE. The interaction of NO with prostaglandins released from the EE was investigated by administration of substance P to isolated papillary muscles with an intact EE incubated with arachidonic acid, or with indomethacin. In the second group, the isolated muscles were incubated with the NO synthase inhibitor, N_G-nitro-L-arginine (L-NOARG) during the stabilization after dissection from the heart. This was done to ensure the suppression of NO release. These muscles were subsequently incubated with arachidonic acid or indomethacin. The two groups of muscles thus provide information regarding interaction of NO and prostaglandins when their release was either maximally stimulated or well suppressed. The results demonstrated that the stimulation of endogenous NO and prostaglandin (prostacyclin and prostaglandin E_2) production resulted in a qualitatively different contractile response to one factor depending on the stimulation or inhibition of the release of the other factor from the EE. Stimulation of endogenous NO release by substance P caused a negative inotropic effect with reduction of twitch duration which was preserved in presence of the cyclo-oxygenase inhibitor, indomethacin. However, in the presence of arachidonic acid, this effect of substance P was not observed (Fig. 5.4). Thus effects of endogenous NO were counteracted by the increased production of prostaglandins by administration of arachidonic acid and preserved when the release of cyclooxygenase metabolites was inhibited by indomethacin. On the other hand, on inhibition of the basal release of NO by incubation with L-NOARG, qualitatively opposite responses on peak active twitch tension and twitch duration of the muscle were observed with stimulation and inhibition of release of prostaglandins from the EE (Fig. 5.4). While administration of arachidonic acid in presence of L-NOARG had a positive inotropic effect with prolongation of twitch duration, indomethacin had an opposite effect. The basal release of prostaglandins from the EE thus appears to prolong twitch duration and may increase peak twitch tension, in particular when concomitant release of NO is suppressed or diminished.

It may be deduced from these experiments that basal release of NO and prostaglandins, similar to the regulation of vascular tone in blood vessels, may maintain myocardial performance in basal conditions. It seems that net balance of NO and prostaglandin effects may determine the basal contractile state of the myocardium; this interaction may reflect the interaction of the different second messenger systems in regulation of myocardial function. On the basis of these observations we may postulate the existence of a bio-feedback system in the EE which modulates myocardial performance. The postulated existence of a bio-feedback system in the EE supports the sensory function of the EE which was suggested by Brutsaert in earlier experiments on the EE mediated modulation of underlying myocardium. These results may also partly explain the mechanism of the positive inotropic effect induced by the presence of an intact EE. The basal state of

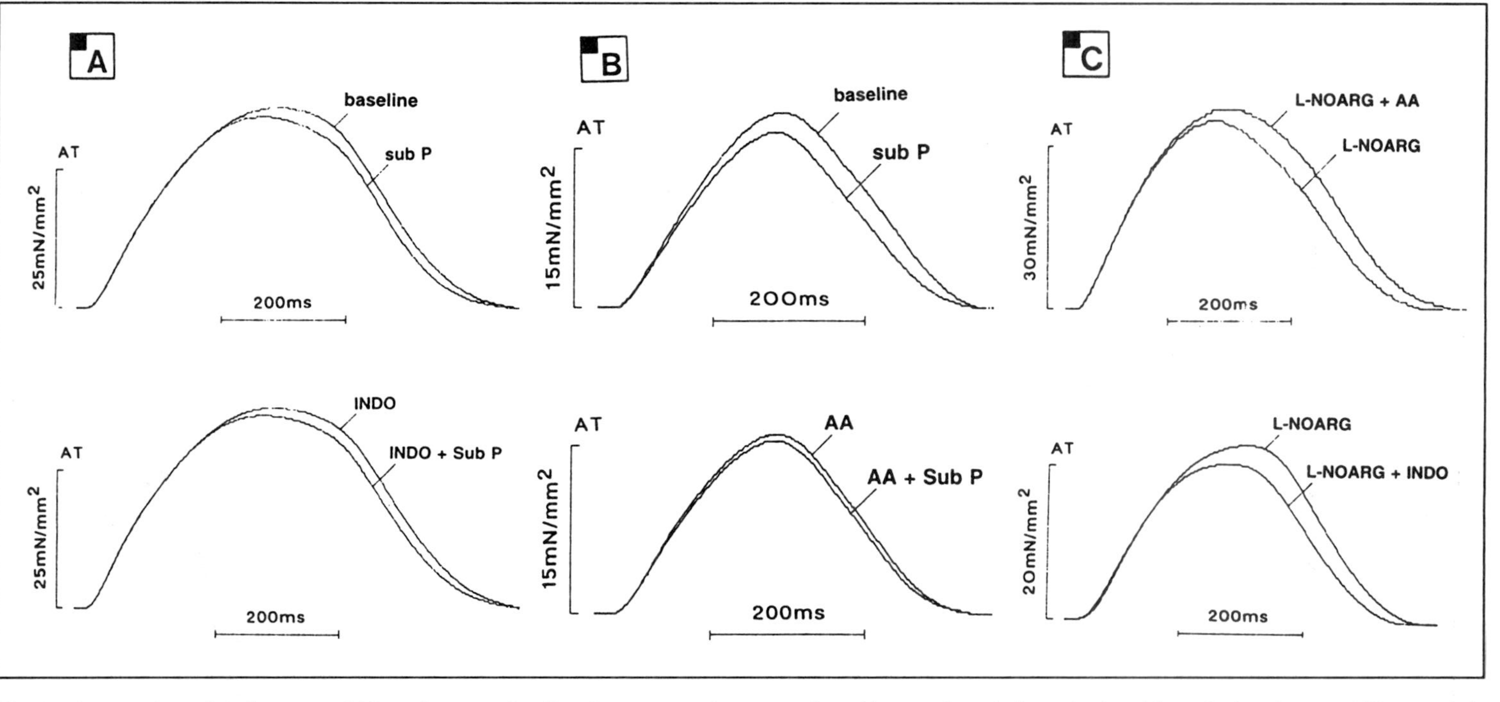

Fig. 5.4. Interaction of endogenous NO and prostaglandins. Representative examples of isometric twitches obtained from isolated cat papillary muscle (Krebs-Ringer; Ca^{2+} 1.25mM; 35°C) showing response to substance P (sub P; 10^{-6} M) in absence (top) and presence (bottom) of indomethacin (INDO; 10^{-6} M) (Panel A) and in absence (top) and presence (bottom) of arachidonic acid (AA; 10^{-5} M) (Panel B). Panel C: representative isometric twitches showing effects of administration of arachidonic acid (AA; 10^{-5} M) and of indomethacin (INDO; 10^{-6} M) in presence of N^G-nitro-L-arginine (L-NOARG) (3×10^{-7} M). AT = peak active isometric tension. Response to substance P was preserved in presence of indomethacin while it was abolished by arachidonic acid. In presence of L-NOARG, arachidonic acid increased AT and prolonged the twitch, in contrast to indomethacin.

myocardium may thus be modulated by the EE by the release of the various biological factors (NO, prostaglandins and endothelin), manifesting their effects through different second messengers and their interactions with each other. The effects of endogenous prostaglandins are only unmasked when concomitant release of endogenous NO is inhibited. The basal release of prostaglandins may thus account for the twitch prolonging effect imparted by the intact EE and may also contribute to the positive inotropic effect. The underlying subcellular mechanism(s) involved in myocardial effects of prostaglandins (Fig. 5.5) still need to be explored and may involve the second messengers, cAMP and IP_3/DAG.

3. ROLE OF ENDOTHELIN

Endothelin is a recently discovered 21-amino-acid peptide released from many endothelial cells. It is one of the most potent vasoconstrictive and cardiotonic peptides known so far and displays a high density of receptors in the myocardium.[8] Although the cardiac pharmacological activities of endothelin are indisputable, the role of endothelin in cardiac physiology remains to be established.

The demonstration of mRNA for endothelin-1 in the EE and its release from EE in culture,[31] release of endothelin-1 from the EE in rat trabeculae[28] and in isolated cardiac muscle[10,55] suggests an important role for endothelin in the EE-mediated regulation of cardiac function. The positive inotropic effect of exogenous endothelin in isolated papillary muscle from various animal species resembles the characteristic inotropic changes induced by the EE.[57] In addition, as with the EE, the positive inotropic activity of endothelin is largely explained by an enhanced responsiveness of the contractile proteins to calcium.[59] Accordingly, endothelin seems to be an important contributor to the positive inotropic effect of intact EE. Moreover, there is growing experimental evidence that endothelin may interact with EE-derived NO and prostaglandins. Interaction between endothelin, NO and prostaglandins has been demonstrated for vascular endothelium and smooth muscle.[26] Only full explanation of this interaction will allow us to understand how local release of endocardial and (micro)vascular endothelial substances influences the performance of underlying myocytes.

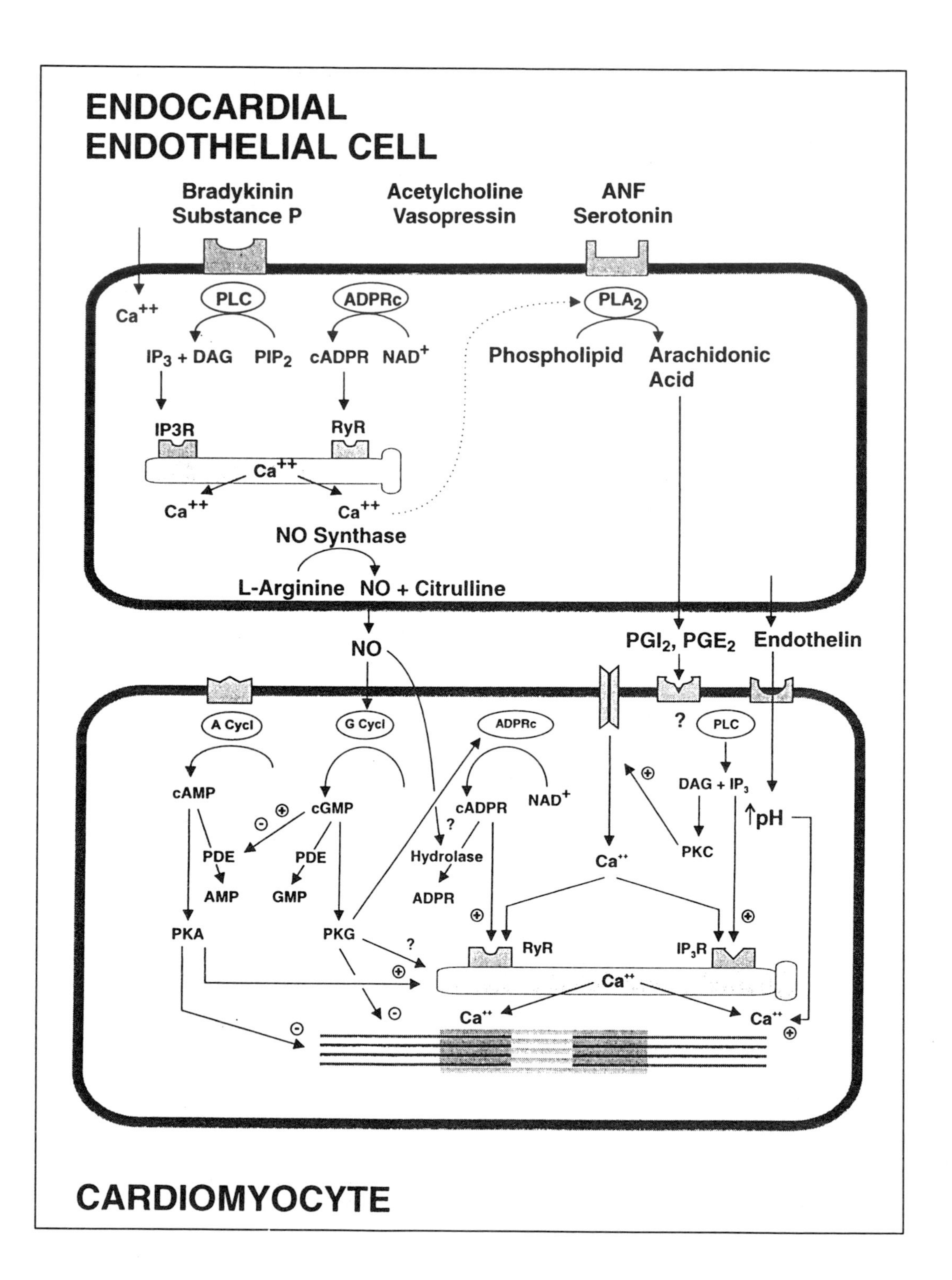
ENDOCARDIAL
ENDOTHELIAL CELL
Bradykinin
Substance P
Acetylcholine
Vasopressin
ANF
Serotonin
Ca++
PLC
ADPRc
PLA2
IP3 + DAG
PIP2
cADPR
NAD+
Phospholipid
Arachidonic
Acid
IP3R
RyR
Ca++
Ca++
Ca++
NO Synthase
L-Arginine
NO + Citrulline
NO
PGI2, PGE2
Endothelin
A Cycl
G Cycl
ADPRc
?
PLC
cAMP
cGMP
cADPR
NAD+
DAG + IP3
↑pH
PDE
PDE
Hydrolase
Ca++
PKC
AMP
GMP
ADPR
PKA
PKG
RyR
IP3R
Ca++
Ca++
Ca++
CARDIOMYOCYTE

A role for endothelin in normal cardiac physiology may appear puzzling. *First*, circulating levels of endothelin in healthy humans (1-5 pM[1] (and even in pathological states (25 pM)) are much lower than the concentrations necessary to occupy endothelin receptors (Kd = 1 nM). This difficulty is usually disproved by the argument that endothelin is preferentially released at the abluminal surface of the endothelial cells and is locally degraded by peptidases in the tissue, resulting in low plasma levels. This argument is not endorsed, however, by recent observations that low concentrations of exogenous endothelin (plasma concentration never exceeding 25 pM in these experiments) are active in vivo.[11] *Second*, endothelin enhances oxygen consumption of myocytes by its potent inotropic properties and decreases oxygen supply by its (in vitro) vasoconstrictive actions. Endothelin would, therefore, seem to be detrimental rather than beneficial for the heart.

Several recent discoveries have, however, extensively changed our insights in the biological role of endothelin. These discoveries may explain many of the above difficulties.

First, Frelin and Guedin[11] accurately explained that most of the current inferences about endothelin start from the assumption that endothelin acts under nonstoichiometric binding conditions (defined by a condition where the receptor concentration in the tissue (Ro) is smaller than the equilibrium dissociation constant of endothelin receptor complexes (Kd)). There is, however, growing evidence that in vivo endothelin binds stoichiometrically to its receptor (defined by the condition Kd<Ro). Under these conditions endothelin immediately binds to its receptor resulting in low free plasma concentrations. Such low concentrations would correspond to the low but active plasma endothelin observed during in vivo experiments. In addition, these binding features predict that at basal

Fig. 5.5 (opposite). Mechanism of endothelial regulation of myocardial performance by signal transduction. EE has been shown to release NO, prostaglandins (PG) and endothelin (ET). There may be continuous basal as well as stimulated release in response to various stimuli. In the cardiomyocyte, the biochemical factors will, through intracellular second messengers, mediate their effects on intracellular release of Ca^{2+} and/or change in myofilament sensitivity to Ca^{2+}. This could explain, to some extent, the regulation of myocardial performance by EE. A Cycl, adenylate cyclase; G Cycl, guanylate cyclase; ADPRc, ADP ribocyclase; PLC, phospholipase C; PLA_2, phospholipase A_2; cAMP, cyclic AMP; cGMP, cyclic GMP; cADPR, cyclic ADP ribose; IP_3, inositol triphosphate; DAG, diacylglycerol; PKA, protein kinase A; PKG, protein kinase G; PKC, protein kinase C; IP_3R, IP_3 receptor; RyR, ryanodine receptor.

rates of endothelin secretion endothelin is more likely to act in an autocrine way i.e. by binding at the endothelin-B receptor on the endothelial surface, thereby dilating the vessel indirectly through NO and prostaglandin release.[26] Only at the higher rates of endothelin secretion, e.g., in ischemia, cardiogenic shock, hypercholesterolemia, or other pathological conditions endothelin would diffuse into the deeper layers and promote direct vasoconstriction.

Second, recent experiments have revealed that endothelin beneficially reversed acidosis-induced negative inotropic and lusitropic effects.[58] These effects were, unlike many other inotropic agents, not accompanied by increased intracellular Ca^{2+}. Therefore, during acidosis locally released endothelin from either EE or from (micro)vascular endothelium may be beneficial for the heart.

Finally, two recent studies described how endothelin opposed the adverse effect of catecholamines on cardiac electrophysiology.[18,40] The authors postulated that enhanced release of endothelin during myocardial ischemia, secreted by coronary artery and EE-cells, act as a local paracrine hormone that directly counters the electrophysiological effects of catecholamines. Endothelin may thus mitigate the risk of life-threatening ventricular arrhythmias in the course of a heart attack.

In conclusion, endothelin is synthesized and released from the EE and coronary endothelium. Basal endothelin release from these cells is likely to participate in the positive inotropic tone elicited by the cardiac endothelium. Recently discovered properties of endothelin revealed that low concentrations of this peptide may have an important protective role in the heart. These beneficial features of cardiac endothelin at low to moderate concentrations contrast with the detrimental effects at higher concentrations.

4. GENERAL CONCLUSION

The mechanisms of EE modulation of myocardial performance are still under investigation and the possibilities include the role of EE as a physico-chemical blood-heart barrier or the release of various chemical messengers by the EE analogous to the vascular endothelium. These include NO, prostacyclin and prostaglandin E_2, endothelin and a "myofilament desensitizing factor". There is evidence to suggest that the unique positive inotropic effect imparted by an intact EE associated with twitch prolongation may be explained, at least partly, by the net effects of basal release of endogenous NO, prostaglandins and endothelin. The relative contribution and release of these mediators from the EE may vary according to the circumstances. EE apparently also functions as a sensory organ and could thereby regulate the release of these mediators individually in order to maintain homeostasis. There is also some emerging evidence from our lab that EE may also mediate its effect through regulation of ionic currents (mainly Cl^-) in the cardiomyocyte (see chapter 4). This mechanism may independently, or synergistically with release of biochemical mediators, also play a significant role in modulation of myocardial performance by EE.

References

1. Battistini B, D'Orlean-Juste P, Sirois P. Endothelins: circulating plasma levels and presence in other biological fluids. Lab Invest 1993; 22(Suppl 8):S321-S324.
2. Beavo JA, Reifsnyder DH. Primary sequence of cyclic nucleotide phosphodiesterase isozymes and the design of selective inhibitors. Trends Pharmacol Sci 1990; 11:150-155.
3. Berridge MJ. Cell signalling: a tale of two messengers. Nature 1993; 365:388-389.
4. Brandt R, Nowak J, Sonnenfeld T. Prostaglandin formation from exogenous precursor in homogenates of human cardiac tissue. Basic Res Cardiol. 1984; 79:135-141.
5. Brutsaert DL, Meulemans AL, Sipido KR et al. Effects of damaging the endocardial surface on the mechanical performance of isolated cardiac muscle. Circ Res 1988; 62:357-366.
6. Brutsaert DL. The endocardium. Annu Rev Physiol 1989; 51:263-273.
7. Cheung Lee H. A signalling pathway involving cyclic ADP-ribose, cGMP, and nitric oxide. News in Physiol Sci 1994; 9:134-137.
8. Davenport AP, Nunez DJ, Hall A et al. Autoradiographical localization of binding sites for porcine (^{125}I)endothelin-1 in humans, pigs, and rats: Functional relevance in humans. J Cardiovasc Pharmacol 1989; 161:1252-1259.
9. Dollinger SJ, Wahler GM. A nitric oxide donor has stimulatory and inhibitory effects on the cardiac calcium current, both of which are inhibited by a G-kinase blocker. Biophysical J 1994; 66:A238.
10. Evans HG, Lewis MJ, Shah AM. Modulation of myocardial relaxation by basal release of endothelin from endocardial endothelium. Cardiovasc Res 1994; 28:1694-1699.
11. Frelin C, Guedin D. Why are circulating concentrations of endothelin so low. Cardiovasc Res 1994; 28:1613-1622.
12. Galione A, Cheung Lee H, Busa WB. Ca^{2+}-induced Ca^{2+} release in

sea urchin egg homogenates: modulation by cyclic ADP-ribose. Science 1991; 253:1143-1146.
13. Galione A, White A, Willmott N et al. cGMP mobilizes intracellular Ca^{2+} in sea urchin eggs by stimulating cyclic ADP-ribose synthesis. Nature 1993; 365:456-459.
14. Galione A. Cyclic ADP-ribose: a new way to control calcium. Science 1993; 259:325-326.
15. Gillebert TC, De Hert SG, Andries LJet al. Intracavitary ultrasound impairs left ventricular performance: presumed role of endocardial endothelium. Am J Physiol 1992; 263:H857-H865.
16. Hasebe N, Shen YT, Vatner SF. Inhibition of endothelium-derived relaxing factor enhances myocardial stunning in conscious dogs. Circulation 1993; 88:2862-2871.
17. Ignarro LJ. Biological actions and properties of endothelium-derived nitric oxide formed and released from artery and vein. Circ Res. 1989; 65:1-21.
18. James AF, Xie L, Fujitani Y et al. Inhibition of cardiac protein kinase A-dependent chloride conductance by endothelin-1. Nature 1994; 370:297-300.
19. Klabunde RE, Ritger RC, Helgren MC. Cardiovascular actions of inhibitors of endothelium-derived relaxing factor (nitric oxide) formation/ release in anaesthetized dogs. Eur J Pharmacol 1991; 199:51-59.
20. Klabunde RE, Ritger RC. NG-Monomethyl-L-arginine (NMA) restores arterial blood pressure but reduces cardiac output in a canine model of endotoxic shock. Biochem Biophys Res Commun 1991; 178:1135-1140.
21. Ku DD, Nelson JM, Caulfield JB et al. Release of endothelium-derived relaxing factors from canine cardiac valves. J Cardiovasc Pharmacol. 1990; 16:212-218.
22. Leite-Moreira AF, Mohan P, Sys SU et al. Myocardial positive inotropic effect of dibutyryl-cyclic GMP in-vivo. Eur Heart J 1994; 15:114.
23. Li K, Rouleau JL, Andries LJ et al. Effect of dysfunctional vascular endothelium on myocardial performance in isolated papillary muscles. Circ Res 1993; 72:768-777.
24. Li K, Stewart DJ, Rouleau JL. Myocardial contractile actions of endothelin-1 in rat and rabbit papillary muscles. Role of endocardial endothelium. Circ Res 1991; 69:301-312.
25. Lüscher TF, Boulanger CM, Yang Z et al. Interactions between endothelium-derived relaxing and contracting factors in health and cardiovascular disease. Circulation 1993; 87(suppl V):V36-V44.
26. Manduteanu I, Popov D, Radu A et al. Calf cardiac valvular endothelial cells in culture: production of glycosaminoglycans, prostacyclin and fibronectin. J Mol Cell Cardiol 1988; 20:103-118.
27. McClellan G, Weisberg A, Rose D et al. Endothelial cell storage and release of endothelin as a cardioregulatory mechanism. Circ

Res 1994; 75:85-96.
28. Mebazaa A, Cherian M, Abraham M et al. Endocardial endothelial prostanoid release responds to flow and hypoxia with response greater than that of the vascular endothelium. Circulation 1993; 88:185.
29. Mebazaa A, Martin LD, Robotham JL et al. Right and left ventricular cultured endocardial endothelium produces prostacyclin and PGE_2. J Mol Cell Cardiol. 1993; 25(3):245-248.
30. Mebazaa A, Mayoux E, Maeda K et al. Paracrine effects of endocardial endothelial cells on myocyte contraction mediated via endothelin. Am J Physiol 265:H1841-H1846.
31. Meulemans AL, Andries LJ, Brutsaert DL. Endocardial endothelium mediates positive inotropic response to alpha1-adrenoreceptor agonist in mammalian heart. J Mol Cell Cardiol 1990; 22:667-685.
32. Meulemans AL, Sipido KR, Sys SU et al. Atriopeptin III induces early relaxation of isolated mammalian papillary muscle. Circ Res 1988; 62:1171-1174.
33. Mohan P, Brutsaert DL, Sys SU. Myocardial performance is modulated by interaction of cardiac endothelium-derived nitric oxide and prostaglandins. Cardiovasc Res 1995; 29:637-640.
34. Mohan P, Brutsaert DL, Sys SU. Inotropic effect of acetylcholine: role of endocardial endothelium. Eur Heart J 1994; 15:283.
35. Mohan P, Sys SU, Brutsaert DL. Nitric oxide donors induce a positive inotropic effect mediated by cGMP in isolated cardiac muscle without endothelium. Eur Heart J 1994; 15:145.
36. Méry P-F, Lohmann SM, Walter U et al. Ca2+ current is regulated by cyclic GMP-dependent protein kinase in mammalian cardiac myocytes. Proc Natl Acad Sci USA 1991; 88: 1197-1201.
37. Méry P-F, Pavoine C, Belhassen L et al. Nitric oxide regulates cardiac Ca^{2+} current. Involvement of cGMP-inhibited and cGMP-stimulated phosphodiesterases through guanyl cyclase activation. J Biol Chem 1993; 268:26286-26295.
38. Mészáros LG, Bak J, Chu A. Cyclic ADP-ribose as an endogenous regulator of the non-skeletal type ryanodine receptor Ca^{2+} channel. Nature 1993; 364:76-79.
39. Ono K, Trautwein W. Potentiation by cyclic GMP of β-adrenergic effect on Ca^{2+} current in guinea-pig ventricular cells. J Physiol (Lond) 1991; 443:387-404.
40. Ono K, Tsujimoto G, Sakamoto A, et al. Endothelin-A receptor mediates cardiac inhibition by regulating calcium and potassium currents. Nature 1994; 370: 301-304.
41. Pfitzer G, Ruegg JC, Flockerzi V et al. cGMP protein kinase decreases calcium sensitivity of skinned cardiac fibres. FEBS Lett. 1982; 149:171-175.
42. Ramaciotti C, McClellan G, Sharkey A et al. Cardiac endothelial cells modulate contractility of rat heart in response to oxygen tension and coronary flow. Circ Res. 1993; 72:1044-1064.

43. Richard V, Berdeaux A, la Rochelle CD et al. Regional coronary hemodynamic effects of two inhibitors of nitric oxide synthesis in anaesthetized, open chest dogs. Br J Pharmacol 1991; 104:59-64.
44. Schoemaker IE, Meulemans AL, Andries LJ et al. Role of the endocardial endothelium in the positive inotropic action of vasopressin. Am J Physiol 1990; 259:H1148-H1151.
45. Schulz R, Smith JA, Lewis MJ et al. Nitric oxide synthase in cultured endocardial cells of the pig. Br J Pharmacol 1991; 104:21-24.
46. Shah AM, Andries LJ, Meulemans AL et al. Endocardium modulates inotropic response to 5-hydroxytryptamine. Am J Physiol 1989; 257:H1790-H1797.
47. Shah AM, Lewis MJ, Henderson AH. Effects of 8-bromo-cyclic GMP on contraction and on inotropic response of ferret cardiac muscle. J Mol Cell Cardiol 1991; 23:55-64.
48. Shah AM, Mebazaa A, Wetzel RC et al. Novel cardiac myofilament desensitizing factor released by endocardial and vascular endothelial cells. Circulation 1994; 89:2492-2497.
49. Shah AM, Meulemans AL, Brutsaert DL. Myocardial inotropic responses to aggregating platelets and modulation by the endocardium. Circulation 1989; 79:1319-1323.
50. Shah AM, Smith JA, Lewis MJ. The role of endocardium in the modulation of contraction of isolated papillary muscles of the ferret. J Cardiovasc Pharmacol 1991; 17 (S3):S251-S257.
51. Shah AM, Spurgeon HA, Sollot SJ et al. 8-bromo-cGMP reduces the myofilament response to Ca^{2+} in intact cardiac myocytes. Circ Res 1994; 74:970-978.
52. Smith JA, Shah AM, Lewis MJ. Factors released from endocardium of the ferret and pig modulate myocardial contraction. J Physiol 1991; 439:1-14.
53. Stamler JS, Loh E, Roddy MA et al. Nitric oxide regulates basal systemic and pulmonary vascular resistance in healthy humans. Circulation 1994; 89:2035-2040.
54. Takanashi M, Endoh M. Characterization of positive inotropic effect of endothelin on mammalian ventricular myocardium. Am J Physiol 1991; 261:H611-H619.
55. Walter U. Physiological role of cGMP and cGMP-dependent protein kinase in the cardiovascular system. Rev. Physiol. Biochem. Pharmacol. 1989; 113:42-48.
56. Wang J, Morgan JP. Endocardial endothelium modulates myofilament Ca^{2+} responsiveness in aequorin-loaded ferret myocardium. Circ Res 1992; 70:754-760.
57. Wang J, Morgan JP. Endothelin reverses the effects of acidosis on the intracellular Ca^{2+} transient and contractility in ferret myocardium. Circ Res 1992; 71:631-639.
58. Wang J, Paik G, Morgan J. Endothelin enhances myofilament Ca^{2+} responsiveness in aequorin-loaded ferret myocardium. Circ Res 1991; 69:582-589.

CHAPTER 6

Endocardial Endothelial Dysfunction

Stanislas U. Sys, Gilles De Keulenaer, Grzegorz Kaluza, Luc J. Andries and Dirk L. Brutsaert

"...les maladies de cet élément anatomique du cœur sont beaucoup plus communes qu'on ne l'avait cru jusqu'ici...Elles sont presque toujours méconnues à l'état aigu, et passent souvent à l'état chronique...les affections de l'endocarde...sont le point de départ le plus ordinaire, la cause-mère ou génératrice la plus fréquente de ces nombreuses lésions organiques, soit des valvules, soit des parois et des cavités du cœur..."‡

Bouillaud (1836)

The endocardium is a highly specialized and selectively permeable barrier which separates the circulating blood from the myocardium at the cavitary side of the heart. The endocardial endothelium (EE) modulates mechanical performance of the subjacent myocardium (chapter 3) through its electrophysiological properties (chapter 4) and through the release of a number of substances (chapter 5). In addition the EE may play a role in inflammatory

‡ *"diseases of the endocardium are much more common than anyone had realized up to now; ... endocardial diseases almost always go unrecognized in the acute stage and frequently become chronic; ... diseases of the endocardium are the usual site or origin and the most frequent fundamental or generating cause of the many organic lesions of the valves, walls, or cavities of the heart ... !"*

Endocardial Endothelium: Control of Cardiac Performance, edited by Stanislas U. Sys and Dirk L. Brutsaert. © 1995 R.G. Landes Company.

or immunologic reactions by the expression of adhesion receptors and antigens of the major histocompatibility complex, in coagulant and thrombotic processes and finally in cardiac remodeling through production of growth-promoting or -inhibiting factors. A failure of the EE to fulfill any of the above mentioned functions may be considered as EE dysfunction. This dysfunction can range from insufficient or excessive performance of any function or of several functions to the extreme of loss of endothelial morphological — cytoskeletal or cellular — integrity and frank denudation of the underlying basal lamina, subendothelial fibrous tissue or even myocardium. The introductory statement from Bouillaud[15] will probably find a much broader field of applicability than the field of endocarditis for which it was meant back in 1836. In this chapter, we will review present knowledge on the existence of functionally significant regional or generalized EE dysfunction with or without morphological abnormalities. We will further address arguments for a possible involvement of EE dysfunction in the pathogenesis of cardiac disease either as a primary cause of disease or acting as a deteriorating positive feedback secondary to diseased myocardium.

1. PATHOPHYSIOLOGICAL ROLE OF ENDOCARDIAL ENDOTHELIUM

In chapter 3, we have seen that experimental impairment of EE by Triton immersion, continuous wave ultrasound or air drying importantly affected mechanical performance of the underlying myocardium. Hence EE dysfunction may be expected to have severe consequences in cardiac diseases (Fig. 6.1). The post-mortem pathology of endocardial lesions induced by EE-related diseases has been well described.[53] These diseases include catecholamine-induced endocardial and subendocardial injuries; various forms of parietal endocarditis; EE alteration and intraventricular thrombi in myocardial infarction, in ventricular aneurysms and in dilated cardiomyopathy; hypereosinophilic endomyocardial fibrosis (Loeffler's endocarditis); endocardial fibroelastosis; serotonin-induced right ventricular endocardial fibrosis in the carcinoid syndrome; endocardial changes with advancing age; and several other forms.

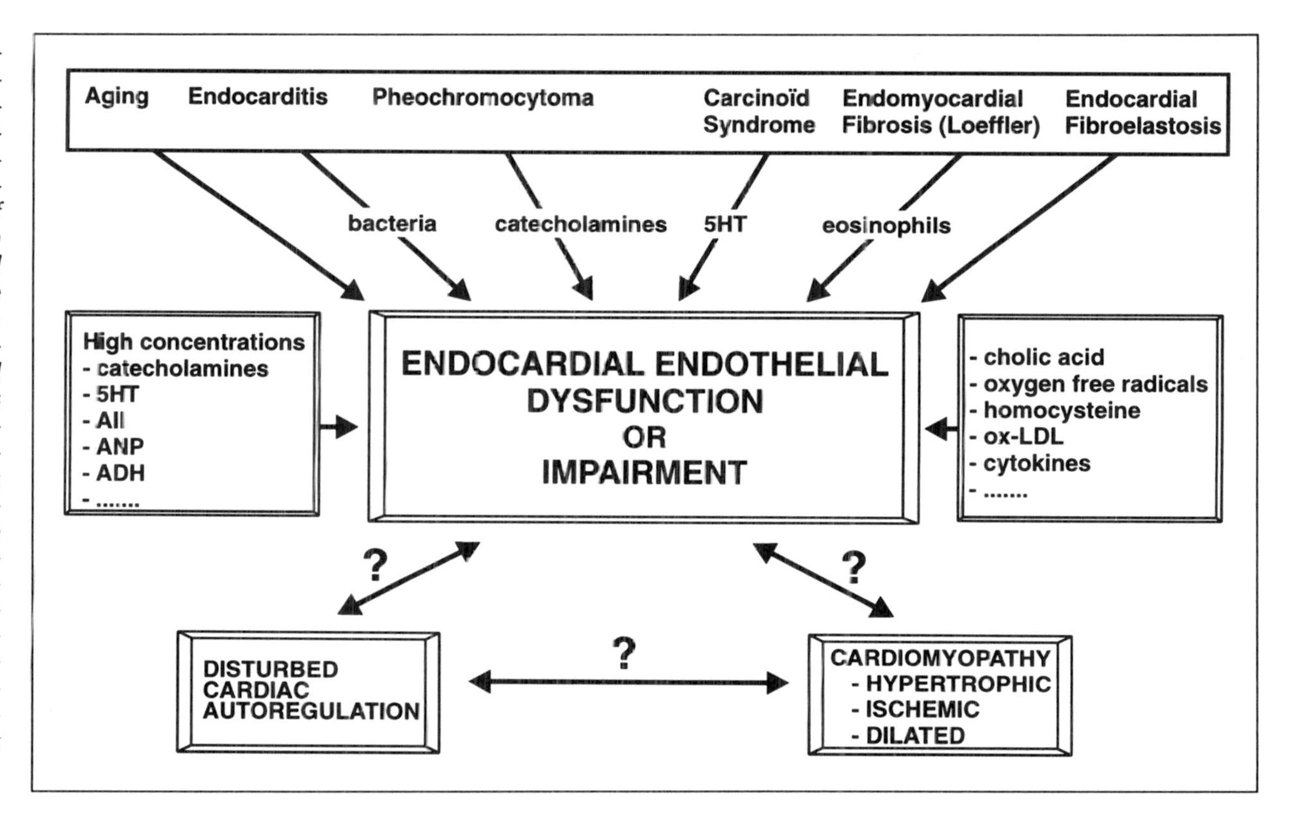

Fig. 6.1. Endocardial endothelial dysfunction: etiologies and possible implications in cardiac failure. Endocardial endothelial lesions are associated with a number of cardiac diseases of which an incomplete list and possible mediators are given on top. In addition, several genuine damaging substances (left) and cardiovascular risk factors (right) have been demonstrated to induce damage to EE cells in vitro; the possibility of protection against some of these factors has been demonstrated. Given the dramatic effect of EE impairment on mechanical performance of myocardium, endocardial damage could profoundly influence the development of cardiac disease.

1.1 Catecholamines and the Endocardium

In patients with pheochromocytomas[3,76,144,152] or in animals after prolonged infusions of catecholamines,[124,146,161] circulating catecholamines are known to induce lesions to the myocardium. Although these lesions develop preferentially in the subendocardial region,[31,71,112,118,120,145] sometimes accompanied by extensive endocardial hemorrhage and edema,[105] damage to the EE has received less attention. If such damage occurs, it could contribute to some of the catecholamine-induced cardiac pathology, including cardiac failure after injection of large doses of epinephrine and in cases of pheochromocytoma.[118]

Two to four hours after acute administration of high dosages of isoproterenol (42 mg/kg, s.c.) in rats, the EE cells showed marked focal degenerative changes with discontinuities near intercellular junctions.[16] In a similar study,[107] endocardial damage was well defined by 30 minutes after acute isoproterenol (10 mg/kg) administration, with contraction and separation of EE cells and with adhesion of activated platelets to the exposed subendocardium. Subendothelial myocardial cellular changes were present at 1 hour only, when the endothelial cell membranes already manifested numerous small pores. In addition, endocardial damage also always preceded leukocytic invasion.[16,107] Hence, endocardial alterations appeared to be the direct result of isoproterenol toxicity rather than a secondary inflammatory response to myocardial necrosis typical for this cardiotoxicity.[32,103,124] After 10 daily dosages of 0.125 mg/kg isoproterenol, an important proliferation of fibroblasts was observed with only minimal injury or necrosis of myocytes; fibroblasts proliferated preferentially in the subendocardial region of the septum.[11] Although this fibrosis was observed in the left and right ventricle and on the rather smooth less-trabeculated septal surface, the spongious myocardium has been reported to be more vulnerable to damage by isoproterenol.[110]

The mechanisms underlying catecholamine-induced injury to the endocardium and myocardium are as yet unknown.[123] As isoproterenol undergoes oxidation with formation of oxygen free radicals, endocardial and subsequent myocardial injury has been attributed to lipid peroxidation.[43] Free radicals have a very short life, yet sufficiently long to initiate lipid peroxidation of circulating lipoproteins. Injury by free radicals and lipid peroxides to endothelial cells has indeed been demonstrated. However, we have not been able to induce EE damage with several oxidation products of catecholamines in the rat in vivo (unpublished results). Alternatively, the presence of adrenoceptors on the EE cells and the shape changes by contractile filaments could be compatible with the view that these cells, as was shown at high concentrations of the α_1-agonist phenylephrine in vitro, could undergo irreversible shape deformation akin to a state of hypercontracture in myocytes.[98] It might be tempting to ascribe catecholamine-induced myocardial hypertrophy too, even in the absence of major hemodynamic overloading,[79] to catecholamine-induced alterations of the EE with subsequent direct exposure of the subjacent myocardium either to catecholamines or to various other substances circulating in the superfusing blood, such as an endothelium-derived growth factor[61,94,136] or an endothelium-derived myocardial contraction factor (MCF).[26] A functional effect of the already mentioned subendocardial proliferation of fibroblasts in the genesis of myocardial hypertrophy after small doses of isoproterenol has also not been excluded. Similar alterations of the EE could perhaps also explain the marked atrial myocardial hypertrophy in various forms of endocarditis.[67] Such adaptations would be akin to the occurrence of focal hypertrophy of vascular smooth muscle after experimental injury to the overlying endothelial cells.[116]

1.2 Myocardial Infarction and the Endocardium

Extensive endocardial lesions have been observed as a consequence of myocardial infarction. In dog hearts, desquamation of the EE cells overlying the ischemic muscle was noted as early as 20 minutes after ligation of a coronary artery, with separation of adjacent endothelial cells along their intercellular boundaries and with loss of many of these cells.[62] Up to 2 hours after coronary ligation the denuded areas from which cells had been lost were smooth, rounded and separated by prominent intercellular lines. After 6 hours, these denuded areas were rough with loss of the basal lamina and the intercellular lines had become indistinct. Subsequent mural thrombosis on the ventricular wall appeared to be initiated by the loss of the basal lamina allowing direct contact between platelets and denuded endocardial collagen, rather than by the loss of endothelial cells per se. When the endocardial lesions were studied after longer periods of time, i.e. at 24 and 48 hours, endothelial desquamation appeared to be the mere secondary result of the myocardial inflammatory response to the infarction with leukocyte invasion separating the endothelial cells from the basal lamina.[72] These cells desquamated in sheets, leaving behind denuded basal lamina, but thrombus formation was rarely observed before 48 hours. Yet, the hypothesis that endocardial damage is a mere consequence of leucocyte invasion is unlikely; endocardial damage has indeed been observed in the absence of leukocytes as early as 20 minutes after coronary ligation in the dog[62] and following 2 hours of global ischemia in the rat.[30] Accordingly, EE lesions in infarction or ischemia seem somehow to result from a primary and early injury to the endothelial cells.

1.3 Infectious Diseases and the Endocardium

Apart from the endothelial cells overlying cardiac valves and atrial myocardium[137,158] less attention has been paid to the effect of infecting agents on the parietal endocardium, in particular the EE cells, overlying ventricular myocardium.[12] Trauma of the endothelial surface of the endocardium by mechanical stress or by turbulence or jet effects has been suggested as a critical factor in the location and evolution of most forms of endocarditis.[80,122,143] Infective endocarditis is believed to be preceded almost certainly by nonbacterial platelet-fibrin thrombotic vegetations.[5] These sterile fibrin-platelet nidi serve to trap organisms, as colonization of these sterile vegetations starts within minutes of the arrival of circulating bacteria.[45] On the other hand, the surface fibrin layer may, in some cases, protect growing colonies from the action of antibiotics.[45-47] The histopathology of the endocardium proper in various forms of endocarditis,[67] including endocarditis parietalis diffusa chronica,[56] has been well established. It consists of edema, infiltration with inflammatory cells, increase in number of smooth muscle cells, vascularization with penetration of capillaries as far as the most inner luminal third of the endocardium proper, and the formation, just beneath the basement membrane, of a new layer consisting of pluripotent mesenchymal cells as the first sign of healing.[67]

In some other infectious diseases, the endothelial surface may also often be seriously damaged. Damage may be caused directly through the infecting agent with considerable shrinkage of individual cells, retracting cell junctions and small holes in the cell membrane.[158] Damage may also result indirectly from the release of toxins. For example, Hubschmann noted in 1917 that endocardial thickening was not uncommonly seen in the dilated hearts of diphtheria.[69] More recently, electron-microscopic studies have shown that endotoxin shock, induced by *Escherichia coli* endotoxin in ponies, caused denudation in both ventricles which ranged from focal areas of a few cells with separation of cell junctions to large areas of denuded basal lamina with adhering leukocytes and platelets.[150] Microvilli were less abundant on the surface of EE cells along the wounded areas.

Septic shock is reportedly associated with significant myocardial depression.[1,14,111] This depression was recently shown to be caused by endotoxin-induced release of cytokines, which in turn induce the inducible nitric oxide synthase.[8,18,55,128] The pathogenesis of endocarditis differs from that of septic shock. The microorganisms involved are mostly gram positive and are present in higher concentrations and in closer contact to EE and myocardium. *Streptococcus faecalis* induced severe damage to the EE at high concentrations in vitro but at the same time caused a marked myocardial positive inotropic effect which was only partly β-adrenoceptor mediated.[127]

1.4 Endomyocardial Fibrosis and the Endocardium

EE cells probably also play a role in the pathogenesis[20,39] of endomyocardial fibrosis. Endomyocardial fibrosis, or endomyocardopathy, has been attributed to parasitic,[49,70,135] viral,[134] immunologic, hypereosinophilic,[83,108,109,140] geochemical,[151] and other factors such as pregnancy or puerperium.[92,96]

Hypereosinophilic endomyocardial fibrosis (Loeffler's endocarditis) is accompanied by marked endocardial fibrous thickening and mural thrombosis.[52,83,108,109] The typical endocardial thickening is characterized by cellular infiltration with macrophages, plasma cells, lymphocytes and mast cells, and specifically by the presence of cell debris of degranulated eosinophils and pyknosis of the EE cell cytoplasma.[130] The eosinophil-associated endocardial injury seems to result in thrombus formation on endocardial surfaces of both ventricles and in endomyocardial fibrosis by subsequent conversion to scar tissue and incorporation in the heart wall.[66,106]

Why the endocardium is so susceptible to eosinophil-dependent injury and what determines the rate at which endocardial disease progresses to the late fibrotic stage is still under investigation.[52,140,141] Endomyocardial biopsy findings showed that the EE cells were the primary targets of the disease process.[52] Alterations to myocytes were minimal and were the consequence of proliferation of fibrous tissue with subsequent separation of the muscle cells. Fibrosis was associated with early alterations in endothelial cells showing evidence of damage as well as regeneration. This early damage of the endothelial cells initiates mural thrombosis, endocardial fibrous thickening and eventually the full-blown endomyocardopathy. Although the disease may evolve either as a restrictive or, though less frequently, as a dilated cardiomyopathy, the contribution, to the pathogenesis of cardiac failure, of a damaged EE has not yet been established.

Our observations on the functional role of the EE cells on myocardial performance, suggest that damage of these cells may play a key role in the development of manifest cardiac failure. Selective injury to the endocardium has been related to the endocardial concentration of toxic eosinophil granule basic proteins.[64,140,155] Hypodense (or degranulated) eosinophils, which have enhanced cytotoxic activity compared to normodense eosinophils,[117] have been found in the peripheral blood of patients with the idiopathic hypereosinophilic syndrome.[113] Moreover, raised serum levels of eosinophil granule proteins have been found in patients with eosinophilic endomyocardial fibrosis.[109,155] Interestingly, vascular endothelial cells may release cytokines which induce cocultered human normodense eosinophils to become hypodense with increased functional capacity.[125] Toxicity of isolated eosinophil granule basic proteins was demonstrated in vitro in isolated cardiac myocytes.[109,147,159] A direct toxic effect of these proteins on EE, impairing thrombomodulin function, may explain the pronounced thromboembolic diathesis which characterizes eosinophilic endocarditis.[139] In addition, eosinophils may damage the endothelium by the release of different components of cytoplasmic granules.[132,138]

Tropical endomyocardial fibrosis is a similar, but more slowly progressing, form of eosinophilic endomyocardial disease.[20,37,38] There is now growing consensus that endomyocardial fibrosis and Loeffler's endocarditis are essentially the same disease simply seen at different stages of development, hence with varying degrees of eosinophilia, endocardial scarring and mural thrombosis.[52,109] According to others however, although chronic parasitic infection may lead to severe endomyocardial fibrosis, eosinophils were found to play no role.[50] Experimental murine malaria causes progressive endocardial edema and extensive endocardial thrombosis.[49] Macrophages may play an important role as early changes included sticking of macrophages to the EE cells and migration to subendothelial areas. This was associated with disruption and leakage of the endothelial monolayer and with subsequent subendothelial edema and infiltrates of lymphocytes, neutrophilic granulocytes, macrophages and fibroblasts. Moreover, these endocardial lesions were plugged by numerous microthrombi which eventually grew out as manifest mural thrombi.

A distinct entity is the endocardial fibroelastosis of infants with marked proliferation of subendocardial fibrous elastic tissue.[13,39] This clinical entity has been associated with a variety of congenital cardiac defects where the fibrous reaction is probably induced by abnormal eddy or jet effects or by cardiac dilatation. Endocardial fibroelastosis has also been associated with a wide variety of metabolic defects or unexplained familial diseases, but the pathogenesis, in particular the role of EE, remains obscure.

1.5 Mural Thrombi and the Endocardium

Both in the old and more recent literature,[9,10,20,39,52,68] it has been emphasized that endocardial and subendocardial pathology is at one stage or another almost always associated with the development of manifest mural thrombi. Most echocardiographers are familiar with the observation, in various cardiac diseases and regardless of the treatment, of spontaneous variations in shape and mobility patterns or of the sudden disappearance of often important mural thrombi.[6,7,17,42,44,95,115,119] Similar to vascular endothelial cells,[87,154] the EE cells have strong antithrombotic properties.[19] Hence, regardless of the etiology of the disease, any lesion of the EE cells might contribute to the thrombotic tendency. Mural thrombus formation could be mediated through increased access to the prothrombotic subendothelium and through reduced fibrinolysis. Yet fibrin thrombi do not form on healthy endothelium which has been traumatized.[114] This feature further emphasizes the powerful anticoagulant and fibrinolytic properties of the normal endocardium.[9]

1.6 Endocardial Tumors

Tumors metastatic to the heart are several times more common than primary tumors.[35] The endocardium or the cardiac valves are rarely involved. In absolute numbers, cardiac metastases are most common in carcinoma of breast and lung.

Although cardiac myxomas are rare, they represent the most common type of primary cardiac tumor. Uncertainty about the histogenesis of intracardiac myxoma partly arose from difficulty in identifying the cell type(s); several possibilities have been suggested.[73] An endocardial or endothelial origin was favored because of the alkaline phosphatase content and other histologic findings.[54,97] Besides smooth muscle cells and fibrocytes, an endothelial cell type was described[142] and was even predominant in a case report.[73] Intense reactivity for factor VIII related antigen as a marker for endothelial cells (and megakaryocytes) was found in the 'lepedic' cells of each of 18 cardiac myxomas in an immunocytochemical study.[102] Cardiac autoreactivity study in patients with myxoma frequently demonstrated anti-endothelial antibodies of the IgG type.[86] By contrast, specific markers for endothelial cells were restricted to either the capillaries[36] or the tumor surface[78] in 11 and 24 myxomas respectively. Therefore, EE cells appear to be involved in at least a number of cardiac myxomas.

1.7 Age-Related Changes of the Endocardium

Important histologic changes of the normal human endocardium occur with age.[4,67,91] In early development, the primitive EE cells undergo striking divergent cytodifferentiation in various regions of the fetal heart;[88] changes of the EE cells with age after birth have not been described. Subjacent to the EE cells, the endocardium comprises a basal lamina, a reticular lamina and a fibroelastic layer. The fibro-elastic layer, a matrix formed by collagen and elastic fibers, contains fibroblasts, myofibroblasts and smooth muscle cells which can form a discontinuous layer in atria.[75] In humans, elastic fibers appear focally in the atrial endocardium at the end of the fifth month of embryonic development.[91] Throughout life, the endocardium shows focal zones and diffuse areas of thickening which represent endocardial hypertrophy with proliferation of elastic fibers, collagen bundles and smooth muscle cells.[91] At birth, the abluminal outer third of the endocardium occasionally contains a few smooth muscle cells.[67] The middle third shows a large number of fibroblasts with fairly abundant cytoplasm and somewhat vesicular rounded nuclei. Within a few months after birth, the cytoplasm disappears and the nuclei become denser and elongated. The number of fibroblasts diminishes within the first year of life and becomes sparse from the third decade on. Large mononuclear cells, occasional lymphocytes and rare polymorphonuclear leukocytes may also be seen at birth, but, at about the fourth decade of life, they have practically disappeared from the normal endocardium. At about the end of the first decade, a greater concentration of smooth muscle cells may occasionally be seen. These cells occur in more or less compact masses and become decidedly more conspicuous in the fourth decade, with prominent endocardial hypertrophy particularly in the left ventricle. In the sixth and seventh decade of life a new pattern appears, with collagenous endocardial reduplications, fragmentation of elastic fibers and collagen and elastic fiber replacement of smooth muscle cells.[91] Focal hypertrophy of endocardial smooth muscle and appearance of fat cells occur in the left atrium. The eighth decade is characterized by maximal sclerotic changes in all chambers generally, which would result from prolonged mechanical stresses. The focal zones of hypertrophic endocardial thickening were suggested to result from turbulence or increased flow while mechanical stress in the heart wall would rather induce diffuse endocardial hypertrophy.

1.8 Regeneration of the Endocardium

Regeneration of the EE monolayer is of special interest. The time needed for re-endothelization seems to be related to the extent of denudation. In experiments where only a small area of the surface had been injured, for example following a single gentle longitudinal rub of isolated cardiac muscle,[26] we found that the typical response on myocardial performance, i.e. early tension decline[25] induced by endocardial damage, disappeared again after about 5 to 10 minutes. As this time is too short for true regeneration of cells, the endothelial monolayer appears in some way to be capable of rapid re-sealing or spreading over restricted areas of the denuded endothelial monolayer. The numerous microvilli and other protuberances of the EE cells may constitute quite a substantial membrane reserve which may be utilized for rapid spreading and initiating cell locomotion.[51] Re-endothelization of very small endothelial wounds, e.g., by removal of a single endothelial cell from a confluent monolayer of aortic vascular endothelial cells, was shown to also occur rapidly within 30 to 40 minutes by the extrusion of lamellipodia from all of the adjacent endothelial cells;

actin microfilament bundles which are normally located around the periphery of each endothelial cell are believed to mediate this rapid lamellipodia-mediated wound closure.[157,160]

Although the regenerative power of the endocardium is not yet well understood, EE cells appear to be capable of more rapid re-endothelization than injured vascular endothelial cells which were rehealed only after 2 to 8 days.[114,121] In vascular endothelial regeneration, smooth muscle cells were primarily involved in cellular proliferation in areas which remained denuded after endothelial cell migration.[29,129] EE-cells in culture reached confluence earlier than vascular endothelial cells from aortic and pulmonary artery.[93] The rate of EE cell proliferation was greater and the doubling time for EE cells (34 hours) was shorter than for vascular endothelial cells (45 hours). In the subendothelial endocardium of the ventricles, fibroblasts are more prominent than smooth muscle cells.[16] Subendothelial fibroblasts of the endocardium are able to transform in vivo to myofibroblasts which were claimed to be involved in covering remaining denuded areas. Myofibroblasts proliferate at a rapid rate as compared to smooth muscle cells.[16]

After focal isoproterenol-induced injury, true regeneration of the EE monolayer was completed only after 16 to 24 hours.[16] In experimental myocardial infarction, extensive desquamation of the EE monolayer was observed up to 48 hours after coronary artery ligation.[62,72] After isoproterenol-induced injury, mainly proliferation of myofibroblasts and monocytic cells started 4 hours after injury to cover up the denuded areas in the EE cell layer. Soon junctional complexes, consisting of tight junctions, gap junctions and desmosomes, were formed between these latter cells and the adjacent intact endothelial cells. These latter cells showed an increased number of microvilli. Throughout the course of impairment and regeneration, focal EE thrombosis developed. These thrombotic processes may be a consequence either of the endocardial damage or of the regeneration. Of additional interest is that in some diseases, e.g., hypertension and atherosclerosis, adaptive alterations may occur in vascular aortic endothelial cells with a striking change in actin filament content.[59,60,156,160] EE adaptive alterations in pathophysiology should thus not a priori be excluded (see below).

In conclusion, although there is yet little known about the pathophysiologic role of the endocardium in the etiology and pathogenesis of cardiac diseases, the available studies suggest that in various pathologic conditions the EE cells may be injured. Whether these injuries are primary or secondary to the disease is yet unclear. As we learn more about the EE cells, it may become evident that structural or functional impairment of these cells may result in altered expression of their paracrine activity, permeability, hemostatic and, perhaps, immunologic potential.

2. DAMAGING FACTORS AND PROTECTION OF ENDOCARDIAL ENDOTHELIUM

In experimental in vitro conditions several of the substances which are thought to be involved in EE pathophysiology were already shown to induce damage to the EE. A number of these substances have experimentally been demonstrated to be genuine damaging factors (high concentrations of, e.g., catecholamines,[98] serotonin,[131] angiotensin II,[99] antidiuretic hormone[126] or atrial natriuretic peptide), while other substances are included in any list of cardiovascular risk factors (e.g., cholic acid,[34] oxygen free radicals,[40] homocysteine, oxidized low density lipoproteins or cytokines). On scanning electron microscopic micrographs (SEM), the EE in normal conditions shows centrally placed nuclear bulges, a polygonal pattern of distinct raised intercellular borders and a luminal surface with small microvilli and invaginations.[23,24,98,99,131] The SEM aspect of damaged EE varies between small holes in the membrane, cell retraction with intercellular gaps, severely damaged cell membranes and may go as far as extensive areas devoid of EE cells with remnants of nuclei, cell membranes and cytoskeleton.[98,99,131] Structural features of the intercellular clefts, the glycocalyx and the cytoskeleton are important modulators of endothelial permeability, in particular of paracellular transport (chapter 2), and may be at the origin of changes in the physiological properties of endothelial function. The described changes reveal the underlying basal lamina and expose extracellular material to the superfusing blood, thereby allowing adhesion of platelets and bacteria and infiltration of leukocytes.[148] For the expression of adhesion molecules on the EE membrane surface we refer to chapter 2. These changes obviously also challenge the selective permeability and the barrier function of the EE for various ions or substances between the superfusing blood and the underlying myocardium. Changes in the organization of the cytoskeleton by changes in secondary messengers have been linked to alterations in transcellular permeability in vascular endothelium. Another cause of increased permeability, resulting in interstitial edema and reduced cardiac function, may be the disruption of the glycocalyx of cardiac endothelial cells as described in the rat with hypoxic stress on isolated hearts[153] or with induction of sepsis as a model of multiorgan failure.[65]

High levels of serotonin may be found during platelet aggregation and in conditions of pulmonary embolism, endotoxic shock and the carcinoid syndrome.[2,63,85] A high incidence of echocardiographic abnormalities is reported in carcinoid syndrome.[84] In more severe cases of this condition, high levels of serotonin and other vasoactive substances have also been linked to endocardial fibrosis,[2,149] consisting of abundant acid mucopolysaccharides and collagen but without increase in elastic fibers.[84] Exposure of isolated cat papillary muscle to serotonin resulted in dose-dependent damage to the EE, the severity of which was markedly reduced in presence of the serotonin$_2$-receptor antagonist ketanserin.[131] The question was raised whether ketanserin might reduce the severity of endocardial damage and fibrosis in carcinoid syndrome.

Morphological EE damage at high concentrations of the α_1-agonist phenylephrine, similarly reported for corneal endothelial cell cultures,[4] could be fully prevented by the specific α_1-blocker prazosin.[98] The α_1-receptor stimulation thus appeared to play a significant role possibly through interaction with the mechanisms which regulate cell shape and cell adhesion. Elevated $[Ca^{2+}]_o$ also protected against EE damage induced by the α_1-agonist. There is evidence from vascular endothelial biology that $[Ca^{2+}]_o$ may help to preserve functional and structural integrity of cellular membranes and of intercellular connections;[33,104] moreover, sealing processes in heart tissue are also largely dependent on $[Ca^{2+}]_o$.[41]

In conclusion, several genuine damaging substances or cardiovascular risk factors have been shown to induce more or less selective EE damage in vitro. Although there is yet little known about protection of EE against these noxious factors, in some instances either a specific or a nonspecific protection could be provided by a receptor antagonist or by elevated $[Ca^{2+}]_o$ respectively. Given the recent knowledge regarding vascular endothelial protection, future research on EE will hopefully lead to new interventions to protect EE and thereby cardiac muscle and pump function against pathophysiological conditions.

3. CHRONIC HEART FAILURE AND ENDOCARDIAL ENDOTHELIAL DYSFUNCTION

In most cardiac diseases, regional or global overloading of the ventricle constitutes a common initiator. In subsequent pathogenesis, several compensatory mechanisms are elicited: direct or intrinsic cardiac compensation, in particular dilatation of the ventricle and myocardial hypertrophy, and indirect or extrinsic extracardiac compensation as the result of activation of various hormonal systems. Both cardiac and extracardiac compensatory mechanisms act in concert to compensate initially for the regional or global overloading of the ventricle. These mechanisms may, however, progressively lead to a state of uncontrolled overcompensation, which in turn creates a vicious cycle eventually leading to irreversible chronic congestive heart failure (Fig. 6.2). From our studies on the functional role of EE, one would expect that EE, by virtue of the blood-heart barrier (chapter 4), may participate in these two compensatory mechanisms:[22,24] (i) directly as a modulator of myocardial performance by virtue of the release of cardioactive substances,[26,40,101] along with dilatation and hypertrophy, and (ii) indirectly as a modulator or as the exclusive mediator of the inotropic action of several substances[21,40,98,100,126,133] in the superfusing blood (chapters 3 and 5). Whether EE contributes to the pathogenesis of congestive heart failure — through interaction with the intrinsic or the extrinsic compensatory mechanisms — is still a matter of speculation. We do however have some important arguments which are highly suggestive for such an EE contribution.

There is now compelling experimental evidence that the EE influences myocardial performance through modulation of the responsiveness of the contractile proteins to intracellular calcium, similar to the effects of an increased muscle length (chapter 3). It is conceivable that, in pathological conditions where this influence of EE is diminished, a demand for increased responsiveness to calcium can only be met with an increase in myocardial fiber length, thereby initiating the development of ventricular dilatation.

On the other hand, it is noteworthy that the presence of EE cells results in the retainment of the adult phenotype for myocytes

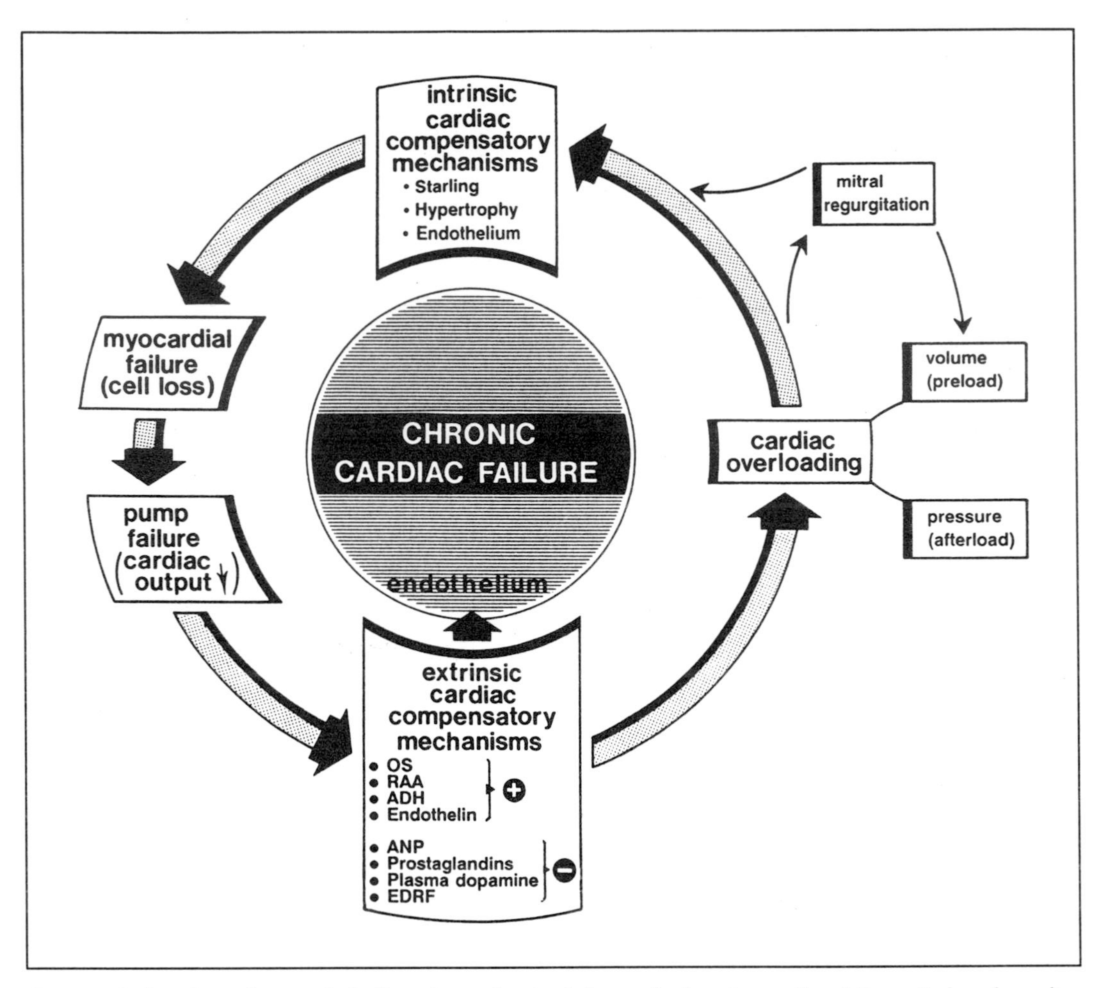

Fig. 6.2. Role of cardiac endothelium in pathophysiology of chronic cardiac failure. Role of cardiac endothelium in the pathophysiology of intrinsic and extrinsic cardiac compensatory mechanisms in the pathogenesis of chronic cardiac failure. OS, orthosympathetic system; RAA, renin-angiotensin-aldosterone; ADH, antidiuretic hormone; ANP, atrial natriuretic peptide. Modified after Brutsaert[22] and Brutsaert et al.[24]

in culture,[48,74,77] suggesting that the EE contributes to the control of cellular differentiation and growth in the myocardium. Interestingly, endocardium of the left ventricle in thechronically volume-overloaded canine heart was significantly thicker than in control hearts, with elastofibrosis and thickening of the basal lamina and with EE cells characterized by the appearance of microfilament bundles, a decreased number of pinocytotic vesicles and an increased number of microvilli.[90]

Recent in vitro and in vivo experiments have shown that the EE mediates the myocardial responsiveness the α_1-adrenergic receptor agonist phenylephrine in the bathing solution or superfusing blood.[81,98] In another study of the pacing-induced model of congestive heart failure in the dog, a marked desensitization to the positive inotropic effects of phenylephrine was observed.[28] In search for a role for the EE in this desensitization of the failing heart, the authors subsequently[82] showed that the EE appeared morphologically intact, that the EE continued to modulate myocardial performance in the previously described manner and that the decrease in α_1-adrenergic responsiveness was indeed EE-dependent. Along the same line, we have recently adapted a method[57,58,89] to obtain pacing-induced heart failure in the rabbit by using an incremental protocol of pacing (unpublished data). After 22 to 26 days of pacing, fully symptomatic heart failure occurred leading to death within 24 to 48 hours. The left ventricles of dilated rabbit hearts were processed for *en face* confocal microscopy. In the apex, EE cells contained abundant actin filament bundles, while peripheral actin bands could either not be discerned or were broad and consisted of disconnected actin filament bundles. In the fibro-elastic layer subjacent to EE cells with a modified actin cytoskeleton, more actin-rich interstitial cells — fibroblasts — were observed than in control hearts. The morphological changes in the EE cells, probably resulting from disturbed mechanical stress on the EE in the heart wall, might again be responsible for the induction of the proliferation of subendothelial cells. It still remains speculative, however, whether these observations can be generalized to other models of congestive heart failure.

Oxygen free radicals are at least partly responsible for endothelial dysfunction in several cardiovascular diseases. We studied the influence of electrolysis-generated oxygen free radicals on morphology and functional interaction of EE and myocardium in isolated cardiac muscle.[40] A brief exposure of the muscle to electrolysis-generated oxygen free radicals increased contractile performance and at the same time partly damaged the EE surface. Because the positive inotropic effect only occurred in the presence of a previously intact EE and was suppressed by preincubation with BQ123, an endothelin-A receptor antagonist, endothelin release from the endothelial cells was proposed to be responsible for the positive inotropic effect. Therefore we concluded that endothelin release from a dysfunctional cardiac endothelium, induced by oxygen free radicals, may indeed be involved in the progression of cardiac disease by influencing myocardial performance in addition to coronary tone and cardiac cell growth.

4. GENERAL CONCLUSION

Various pathologic conditions in cardiovascular biology seem to involve the endothelial and subendothelial endocardial cells. Structural or functional impairment of these cells may be primary or secondary to the disease and may result in altered expression of their paracrine activity, permeability, hemostatic or immunologic potential. Most importantly, however, given the dramatic effect of endocardial impairment on mechanical performance of the subjacent myocardium, one ought not be surprised that endocardial damage could profoundly influence the further development of any cardiac disease. Hence, impaired EE and EE protection could probably play a pivotal role in the etiology/pathogenesis and prophylaxis/treatment of heart failure in various cardiac diseases.

REFERENCES

1. Abel FL. Does the heart fail in endotoxin shock? Circ Shock 1990; 30:5-13.
2. Ahlman A. Serotonin and the carcinoid syndrome. In: Vanhoutte PM, ed. Serotonin and the cardiovascular system. New York: Raven, 1958:199-212.
3. Alpert LI, Pai SH, Zak FG et al. Cardiomyopathy associated with a pheochromocytoma. Report of a case with ultrastructural examination of the myocardial lesions. Arch Pathol 1972; 93:544-548.
4. Andries LJ, Sys SU and Brutsaert DL. Morphoregulatory interactions of endocardial endothelium and extracellular material in the heart. Herz 1995; 20:135-145.
5. Angrist A and Oka M. Pathogenesis of bacterial endocarditis. J Am Med Assoc 1963; 183:117-120.
6. Asinger RW, Mikell FL, Elsperger J et al. Incidence of left-ventricular thrombosis after acute transmural myocardial infarction. Serial evaluation by two-dimensional echocardiography. N Engl J Med 1981; 305:297-302.
7. Asinger RW, Mikell FL, Sharma B et al. Observations on detecting left ventricular thrombus with two dimensional echocardiography: emphasis on avoidance of false positive diagnoses. Am J Cardiol 1981; 47:145-156.
8. Balligand JL, Ungureanu D, Kelly RA et al. Abnormal contractile function due to induction of nitric oxide synthesis in rat cardiac

myocytes follows exposure to activated macrophage-conditioned medium. J Clin Invest 1993; 91:2314-2319.
9. Becker BJP. Studies of the human mural endocardium. J Pathol Bacteriol 1964; 88:541-547.
10. Becker BJP, Chatgidakis CB and Van Lingen B. Cardiovascular collagenosis with parietal endocardial thrombosis. Circulation 1953; 7:345-356.
11. Benjamin IJ, Jalil JE, Tan LB et al. Isoproterenol-induced myocardial fibrosis in relation to myocyte necrosis. Circ Res 1989; 65:657-670.
12. Bisno AL, Dismukes WE, Durack DT et al. Antimicrobial treatment of infective endocarditis due to viridans streptococci, enterococci, and staphylococci. JAMA 1989; 261:1471-1477.
13. Black-Schaffer B. Infantile endocardial fibroelastosis. Archives of Pathol 1957; 63:281-306.
14. Bone RC. The pathogenesis of sepsis. Ann Intern Med 1991; 115:457-469.
15. Bouillaud J. Traité clinique des maladies du coeur, précédé de recherches nouvelles sur l'anatomie et la physiologie de cet organe. Dumont H, Bruxelles, 1836:15-16.
16. Boutet M, Turcotte H, Bazin M et al. An ultrastructural study of endocardial endothelium alterations in catecholamine-induced infarct-like necrosis. Rev Can Biol Exp 1983; 42:87-99.
17. Boyd MT, Seward JB, Tajik AJ et al. Frequency and location of prominent left ventricular trabeculations at autopsy in 474 normal human hearts: implications for evaluation of mural thrombi by two-dimensional echocardiography. J Am Coll Cardiol 1987; 9:323-326.
18. Brady AJ, Poole-Wilson PA, Harding SE et al. Nitric oxide production within cardiac myocytes reduces their contractility in endotoxemia. Am J Physiol 1992; 263:H1963-H1966.
19. Brandt JT. The role of natural coagulation inhibitors in hemostasis. Clin Lab Med 1984; 4:245-284.
20. Brockington IF and Olsen EG. Eosinophilia and endomyocardial fibrosis. Postgrad Med J 1972; 48:740-741.
21. Brutsaert DL. The endocardium. Annu Rev Physiol 1989; 51:263-273.
22. Brutsaert DL. Role of endocardium in cardiac overloading and failure. Eur Heart J 1991; 11(suppl G):8-16.
23. Brutsaert DL. Endocardial and coronary endothelial control of cardiac performance. NIPS 1993; 8:82-86.
24. Brutsaert DL and Andries LJ. The endocardial endothelium. Am J Physiol 1992; 263:H985-H1002.
25. Brutsaert DL and Sys SU. Relaxation and diastole of the heart. Physiol Rev 1989; 69:1228-1315.
26. Brutsaert DL, Meulemans AL, Sipido KR et al. Endocardial con-

trol of myocardial performance. Adv Exp Med Biol 1988; 226:609-615.

27. Brutsaert DL, Meulemans AL, Sipido KR et al. Effects of damaging the endocardial surface on the mechanical performance of isolated cardiac muscle. Circ Res 1988; 62:358-366.
28. Calderone A, Bouvier M, Li K et al. Dysfunction of the beta- and alpha-adrenergic systems in a model of congestive heart failure. The pacing-overdrive dog. Circ Res 1991; 69:332-343.
29. Campbell JH and Campbell GR. Endothelial cell influences on vascular smooth muscle phenotype. Ann Rev Physiol 1986; 48:295-306.
30. Carter G and Gavin JB. Morphological changes in endocardium subjected to global ischaemia. Basic Res Cardiol 1986; 81:465-472.
31. Cebelin MS and Hirsch CS. Human stress cardiomyopathy. Myocardial lesions in victims of homicidal assaults without internal injuries. Hum Pathol 1980; 11:123-132.
32. Chappel CI, Rona G, Balazs T et al. Severe myocardial necrosis produced by isoproterenol in the rat. Arch Int Pharmacodyn 1951; 122:123-128.
33. Clementi F and Palade GE. Intestinal capillaries. II. Structural effects of EDTA and histamine. J Cell Biol 1969; 42:706-714.
34. Colpaert CG, Vandenbroucke MP, Andries LJ et al. Role of endocardial endothelium in the positive inotropic effect of cholic acid in isolated myocardium. J Cardiovasc Pharmacol 1992; 20 Suppl 12:S179-182.
35. Colucci WS and Braunwald E. Primary tumors of the heart In: Braunwald E, ed. Heart Disease. 3rd ed. Philadelphia: Saunders, 1988:1470.
36. Curschellas E, Toia D, Borner M et al. Cardiac myxomas: immunohistochemical study of benign and malignant variants. Virchows Arch (A) 1991; 418:485-491.
37. Davies J, Spry CJ, Vijayaraghavan G et al. A comparison of the clinical and cardiological features of endomyocardial disease in temperate and tropical regions. Postgrad Med J 1983; 59:179-185.
38. Davies JNP. Endocardial fibrosis in Africans. East Afr Med J 1948; 25:10.
39. Davies JNP. Some considerations regarding obscure diseases affecting the mural endocardium. Am Heart J 1960; 49:600-631.
40. De Keulenaer GW, Andries LJ, Sys SU et al. Endothelin-mediated positive inotropic effect induced by reactive oxygen species in isolated cardiac muscle. Circ Res 1995; 76:878-884.
41. De Mello WC. Electrical Phenomena in the heart Academic Press, New York, 1972, chapter 2.
42. DeMaria AN, Bommer W, Neumann A et al. Left ventricular thrombi identified by cross-sectional echocardiography. Ann Intern Med 1979; 90:14-18.
43. Dhalla NS, Ganguly PK, Panagia V et al. Catecholamine-induced

cardiomyopathy: alterations in Ca^{2+} transport systems. In: Kawai C and Abelmann WH, ed. Pathogenesis of myocarditis and cardiomyopathy. Recent experimental and clinical studies. Tokyo: University of Tokyo Press, 1987:135-147.
44. Domenicucci S, Bellotti P, Chiarella F et al. Spontaneous morphologic changes in left ventricular thrombi: a prospective two-dimensional echocardiographic study. Circulation 1987; 75:737-743.
45. Durack DT. Experimental bacterial endocarditis. IV. Structure and evolution of very early lesions. J Pathol 1975; 115:81-88.
46. Durack DT and Beeson PB. Experimental bacterial endocarditis. I. Colonization of a sterile vegetation. Br J Exp Pathol 1972; 53:44-49.
47. Durack DT and Beeson PB. Experimental bacterial endocarditis. II. Survival of a bacteria in endocardial vegetations. Br J Exp Pathol 1972; 53:50-53.
48. Eid H, O'Neill M, Smith TW et al. Plasticity of adult rat cardiac myocytes in long-term culture: role of non-muscle cardiocytes in the maintenance of the adult phenotype in vitro. Circulation 1990; 82:III-689.
49. Eling WM, Jerusalem CR and Heinen-Borries U. Role of macrophages in the pathogenesis of endomyocardial fibrosis in murine malaria. Trans R Soc Trop Med Hyg 1984; 78:43-48.
50. Eling WM, Jerusalem CR, Hermsen CC et al. Endomyocardial lesion and endomyocardial fibrosis in experimental malaria (Plasmodium berghei) in mice. Contrib Microbiol Immunol 1983; 7:218-229.
51. Erickson CA and Trinkaus JP. Microvilli and blebs as sources of reserve surface membrane during cell spreading. Exp Cell Res 1976; 99:375-384.
52. Fauci AS, Harley JB, Roberts WC et al. NIH conference. The idiopathic hypereosinophilic syndrome. Clinical, pathophysiologic, and therapeutic considerations. Ann Intern Med 1982; 97:78-92.
53. Ferrans VJ, Butany JW. Ultrastructural pathology of the heart. Diagn Electron 1983; 4:319-473.
54. Fine G, Morales A and HornJr RC. Cardiac myxoma: A morphologic and histogenetic appraisal. Cancer 1968; 22:1156-1162.
55. Finkel MS, Oddis CV, Jacob TD et al. Negative inotropic effects of cytokines on the heart mediated by nitric oxide. Science 1992; 257:387-389.
56. Fossel M. Chronische parietale endocarditis. Beitr Pathol Anat 1942; 107:241-255.
57. Freeman GL and Colston JT. Myocardial depression produced by sustained tachycardia in rabbits. Am J Physiol 1992; 262:H63-H67.
58. Freeman GL and Colston JT. Simple circuit for pacing hearts of experimental animals. Am J Physiol 1992; 262:H1939-H1940.
59. Gabbiani G, Elemer G, Guelpa C et al. Morphologic and functional changes of the aortic intima during experimental hyperten-

sion. Am J Pathol 1979; 96:399-422.
60. Gabbiani G, Gabbiani F, Lombardi D et al. Organization of actin cytoskeleton in normal and regenerating arterial endothelial cells. Proc Natl Acad Sci USA 1983; 80:2361-2364.
61. Gajdusek C, DiCorleto P, Ross R et al. An endothelial cell-derived growth factor. J Cell Biol 1980; 85:467-472.
62. Gavin JB, Wheeler EE and Herdson PB. Scanning electron microscopy of the endocardial endothelium overlying early myocardial infarcts. Pathology 1973; 5:145-148.
63. Gillis CN. Peripheral metabolism of serotonin. In: Vanhoutte PM, ed. Serotonin and the cardiovascular system. New York: Raven, 1985:27-36.
64. Gleich GJ, Frigas E, Loegering DA et al. Cytotoxic properties of the eosinophilic major basic protein. J Immunol 1979; 123: 2925-2927.
65. Gotloib L, Shostak A, Galdi P et al. Loss of microvascular negative charges accompanied by interstitial edema in septic rats' heart. Circ Shock 1992; 36:45-56.
66. Gottdiener JS, Maron BJ, Schooley RT et al. Two-dimensional echocardiographic assessment of the idiopathic hypereosinophilic syndrome. Anatomic basis of mitral regurgitation and peripheral embolization. Circulation 1983; 67:572-578.
67. Gross L. Lesions of the left auricle in rheumatic fever. Am J Pathol 1935; 11:711-743.
68. Hertel MP. Das Verhalten des Endokards bei parietaler Endokarditis und bei allgemeiner Blutdrucksteigerung. Frankfurter Z Pathol 1921; 24:1-57.
69. Hubschmann P. Über Myokarditis und andere pathologisch-anatomische Beobachtungen bei Diphtherie. München Med Wechnschr 1917; 64:73-76.
70. Ive FA, Willis AJ, Ikeme AC et al. Endomyocardial fibrosis and filariasis. Q J Med 1967; 36:495-516.
71. Jalil JE, Doering CW, Janicki JS et al. Fibrillar collagen and myocardial stiffness in the intact hypertrophied rat left ventricle. Circ Res 1989; 64:1041-1050.
72. Johnson RC, Crissman RS and DiDio LJ. Endocardial alterations in myocardial infarction. Lab Invest 1979; 40:183-193.
73. Kelly M and Bhagwat AG. Ultrastructural features of a recurrent endothelial myxoma of the left atrium. Arch Pathol 1972; 93:219-226.
74. Kelly RA, Eid H, Kramer BK et al. Endothelin enhances the contractile responsiveness of adult rat ventricular myocytes to calcium by a pertussis toxin-sensitive pathway. J Clin Invest 1990; 86:1164-1171.
75. Klein W and Böck P. Elastica-positive material in the atrial endocardium. Acta Anat 1983; 116:106-113.
76. Kline IK. Myocardial alterations associated with pheochromo-

cytomas. Am J Pathol 1961; 38:539-551.
77. Kramer BK, Nishida M, Kelly RA et al. Endothelins. Myocardial actions of a new class of cytokines. Circulation 1992; 85:350-356.
78. Krikler DM, Rode J, Davies MJ et al. Atrial myxoma: a tumour in search of its origins. Br Heart J 1992; 67:89-91.
79. Kuribayashi T, Furukawa K, Katsume H et al. Regional differences of myocyte hypertrophy and three-dimensional deformation of the heart. Am J Physiol 1986; 250:H378-H388.
80. Lepeschkin E. On the relation between site of valvular involvement in endocarditis and the blood pressure resting on the valve. Am J Med Sci 1952; 224:318.
81. Li K, Rouleau JL, Andries LJ et al. Effect of dysfunctional vascular endothelium on myocardial performance in isolated papillary muscles. Circ Res 1993; 72:768-777.
82. Li K, Rouleau JL, Calderone A et al. Endocardial function in pacing-induced heart failure in the dog. J Mol Cell Cardiol 1993; 25:529-540.
83. Löffler W. Endocarditis parietalis fibroplastica mit Bluteosinophilie. Schweiz Med Wochenschr 1936; 66:817-820.
84. Lundin L, Funa K, Hansson HE et al. Histochemical and immunohistochemical morphology of carcinoid heart disease. Pathol Res Pract 1991; 187:73-77.
85. Lundin L, Norheim I, Landelius J et al. Carcinoid heart disease: relationship of circulating vasoactive substances to ultrasound-detectable cardiac abnormalities. Circulation 1988; 77:264-269.
86. Maisch B. Immunology of cardiac tumors. Thorac Cardiovasc Surg 1990; 38:157-163.
87. Marcus AJ. Eicosanoid interactions between platelets, endothelial cells, and neutrophils. Methods Enzymol 1990; 187:585-599.
88. Markwald RR, Fitzharris TP and Smith WN. Structural analysis of endocardial cytodifferentiation. Dev Biol 1975; 42:160-180.
89. Masaki H, Imaizumi T, Ando S et al. Production of chronic congestive heart failure by rapid ventricular pacing in the rabbit. Cardiovasc Res 1993; 27:828-831.
90. Masuda H, Kawamura K, Tohda K et al. Endocardium of the left ventricle in volume-loaded canine heart. A histological and ultrastructural study. Acta Pathol Jpn 1989; 39:111-120.
91. McMillan JB and Lev M. The aging heart. I. Endocardium. J Gerontol 1959; 14:268-283.
92. Meadows WR. Idiopathic myocardial failure in the last trimester of pregnancy and the puerperium. Circulation 1957; 15:903-914.
93. Mebazaa A, Wetzel R, Cherian M et al. Comparison between endocardial and great vessel endothelial cells: morphology, growth, and prostaglandin release. Am J Physiol 1995; 87:H250-H259.
94. Meidell RS, Sen A, Henderson SA et al. Alpha 1-adrenergic stimulation of rat myocardial cells increases protein synthesis. Am J Physiol 1986; 251:H1076-H1084.

95. Meltzer RS, Guthaner D, Rakowski H et al. Diagnosis of left ventricular thrombi by two-dimensional echocardiography. Br Heart J 1979; 42:261-265.
96. Melvin JP. Post-portal heart disease. Ann Intern Med 1947; 27:596-609.
97. Merkow LP, Kooros MA, Magovern G et al. Ultrastructure of a cardiac myxoma of the heart: A case report. Arch Pathol 1969; 88:390-398.
98. Meulemans AL, Andries LJ and Brutsaert DL. Endocardial endothelium mediates positive inotropic response to alpha 1-adrenoceptor agonist in mammalian heart. J Mol Cell Cardiol 1990; 22:667-685.
99. Meulemans AL, Andries LJ and Brutsaert DL. Does endocardial endothelium mediate positive inotropic response to angiotensin I and angiotensin II? Circ Res 1990; 66:1591-1601.
100. Meulemans AL, Sipido KR, Sys SU et al. Atriopeptin III induces early relaxation of isolated mammalian papillary muscle. Circ Res 1988; 62:1171-1174.
101. Mohan P, Brutsaert DL and Sys SU. Myocardial performance is modulated by interaction of cardiac endothelium-derived nitric oxide and prostaglandins. Cardiovasc Res 1995; 29:637-640.
102. Morales AR, Fine G, Castro A et al. Cardiac myxoma (endocardioma). An immunocytochemical assessment of histogenesis. Human Pathol 1981; 12:896-899.
103. Moravec J, Hatt PY. Experimental myocardial necrosis induced by isopropylnoradrenaline. Electron microscopic study Pathol Biol 1969; 17:585-595.
104. Nagy Z, Goehlert UG, Wolfe LS et al. Ca2+ depletion-induced disconnection of tight junctions in isolated rat brain microvessels. Acta Neuropathol Berl 1985; 68:48-52.
105. Nahas GG, Manion WC and Brunson JC. Lésions cardiaques par excès de norépinéphrine. La Presse Médicale 1959; 27:1079-1082.
106. Naito H, Nakatsuka M, Yuki K et al. Giant left ventricular thrombi in the hypereosinophilic syndrome: report of two cases J Cardiogr 1986; 16:475-488.
107. Noronha-Dutra AA, Steen EM and Woolf N. The early changes induced by isoproterenol in the endocardium and adjacent myocardium. Am J Pathol 1984; 114:231-239.
108. Olsen EG. Pathological aspects of endomyocardial fibrosis. Postgrad Med J 1983; 59:135-141.
109. Olsen EG and Spry CJ. Relation between eosinophilia and endomyocardial disease. Prog Cardiovasc Dis 1985; 27:241-254.
110. Ostadal B. Phylogenetic and ontogenetic development of the terminal vascular bed in the heart muscle and its effect on the development of experimental cardiac necrosis. II Annual Meeting of the International Study Group for Research in Cardiac Metabolism Istituto Lombardo - Fondazione Baselli Editrice Succ. Fusi - Pavia, 1969:111-132.

111. Parker MM, Shelhamer JH, Bacharach SL et al. Profound but reversible myocardial depression in patients with septic shock. Ann Intern Med 1984; 100:483-490.
112. Pieper GM, Clayton FC, Todd GL et al. Transmural distribution of metabolites and blood flow in the canine left ventricle following isoproterenol infusions. J Pharmacol Exp Ther 1979; 209:334-341.
113. Pincus SH, Schooley WR, DiNapoli AM et al. Metabolic heterogeneity of eosinophils from normal and hypereosinophilic patients. Blood 1981; 58:1175-1185.
114. Poole JCF, Sanders AG, Florey HW. The regeneration of aortic endothelium. J Path Bact 1958; 75:133-143.
115. Ports TA, Cogan J, Schiller NB et al. Echocardiography of left ventricular masses. Circulation 1978; 58:528-536.
116. Potvliege PR and Bourgain RH. The wall reaction to electric micro-injury at branching sites of mesenteric arteries of the rat: an electron microscopic study of intimal cushions. Br J Exp Pathol 1980; 61:324-331.
117. Prin L, Capron M, Tonnel AB et al. Heterogeneity of human peripheral blood eosinophils: variability in cell density and cytotoxic ability in relation to the level and the origin of hypereosinophilia. Int Arch Allergy Appl Immunol 1983; 72:336-346.
118. Raab W. Key position of catecholamines in functional and degenerative cardiovascular pathology. Am J Cardiol 1960; 5:571-578.
119. Reeder GS, Tajik AJ and Seward JB. Left ventricular mural thrombus: two-dimensional echocardiographic diagnosis. Mayo Clin Proc 1981; 56:82-86.
120. Reichenbach DD and Benditt EP. Catecholamines and cardiomyopathy: the pathogenesis and potential importance of myofibrillar degeneration. Hum Pathol 1970; 1:125-150.
121. Reidy MA and Silver M. Endothelial regeneration. VII. Lack of intimal proliferation after defined injury to rat aorta. Am J Pathol 1985; 118:173-177.
122. Rodbard S. Blood velocity and endocarditis. Circulation 1963; 27:18-28.
123. Rona G. Catecholamine cardiotoxicity. J Mol Cell Cardiol 1985; 17:291-306.
124. Rona G, Chappel CI, Balazs T et al. An infarct-like myocardial lesion and other toxic manifestations produced by isoproterenol in the rat. Arch Pathol 1959; 67:443-455.
125. Rothenberg ME, Owen WF Jr, Silberstein DS et al. Eosinophils cocultured with endothelial cells have increased survival and functional properties. Science 1987; 237:645-647.
126. Schoemaker IE, Meulemans AL, Andries LJ et al. Role of endocardial endothelium in positive inotropic action of vasopressin. Am J Physiol 1990; 259:H1148-H1151.
127. Schoemaker IE, Sys SU, Andries LJ et al. Positive inotropic effect of *Streptococcus faecalis* in isolated cardiac muscle. Am J Physiol

1994; 267:H2450-H2461.
128. Schulz R, Nava E and Moncada S. Induction and potential biological relevance of a Ca2+-independent nitric oxide synthase in the myocardium. Br J Pharmacol 1992; 105:575-580.
129. Schwartz SM, Stemerman MB and Benditt EP. The aortic intima. II. Repair of the aortic lining after mechanical denudation. Am J Pathol 1975; 81:15-42.
130. Sekiguchi M, Yu ZX, Take M et al. Ultrastructural features of the endomyocardium in patients with eosinophilic heart disease. An endomyocardial biopsy study. Jpn Circ J 1984; 48:1375-1382.
131. Shah AM, Andries LJ, Meulemans AL et al. Endocardium modulates myocardial inotropic response to 5-hydroxytryptamine. Am J Physiol 1989; 257:H1790-H1797.
132. Shah AM, Brutsaert DL, Meulemans AL et al. Eosinophils from hypereosinophilic patients damage endocardium of isolated feline heart muscle preparations. Circulation 1990; 81:1081-1088.
133. Shah AM, Meulemans AL and Brutsaert DL. Myocardial inotropic responses to aggregating platelets and modulation by the endocardium. Circulation 1989; 79:1315-1323.
134. Shaper AG, Hutt MSR and Coles RM. Endocardial fibrosis. Cardiologia 1968; 52:20ff.
135. Shaper AG, Kaplin MH, Mody NJ et al. Malarial antibodies and autoantibodies to heart and other tissues in the immigrant and indigenous peoples of Uganda. Lancet 1968; 1:1342.
136. Simpson P. Stimulation of hypertrophy of cultured neonatal rat heart cells through an alpha 1-adrenergic receptor and induction of beating through an alpha 1- and beta 1-adrenergic receptor interaction. Evidence for independent regulation of growth and beating. Circ Res 1985; 56:884-894.
137. Sinapius D and Mohring G. Das Endothel der Herzklappen und Vorhöfe im Häutchenpräparat, zugleich ein Beitrag zur Morphologie der Endokarditis. Virchows Arch 1954; 324:588-611.
138. Slungaard A and Mahoney JR. Bromide-dependent toxicity of eosinophil peroxidase for endothelium and isolated working rat hearts: a model for eosinophilic endocarditis. J Exp Med 1991; 173: 117-126.
139. Slungaard A, Vercellotti GM, Tran T et al. Eosinophil cationic granule proteins impair thrombomodulin function. A potential mechanism for thromboembolism in hypereosinophilic heart disease. J Clin Invest 1993; 91:1721-1730.
140. Spry CJ, Davies J, Tai PC et al. Clinical features of fifteen patients with the hypereosinophilic syndrome. Q J Med 1983; 52:1-22.
141. Spry CJF. Eosinophils and endomyocardial fibrosis: A review of clinical and experimental studies, 1980-86. In: Kawai C and Abelmann WH, ed. Pathogenesis of myocarditis and cardiomyopathy. Recent experimental and clinical studies. Tokyo: Uni-

versity of Tokyo Press, 1987:293-310.
142. Stein AA, Mauro J, Thiboden L et al. The histogenesis of cardiac myxomas: relation to other proliferative diseases of subendothelial vaso-form reserve cells. In: Sommers SC, ed. Pathology Annual (Vol.4). New York: Appleton-Century-Crofts, 1969:293-312.
143. Sullam PM, Drake TA and Sande MA. Pathogenesis of endocarditis. Am J Med 1985; 78(6B):110-115.
144. Szakacs J and Fobes CD. l-Norepinephrine. A pathologist's approach to recent investigations. Maryland Med J 1960; 9:83.
145. Szakacs JE and Cannon A. l-Norepinephrine myocarditis. Am J Clin Pathol 1958; 30:425-434.
146. Szakacs JE and Mehlman B. Pathologic changes induced by l-norepinephrine. Am J Cardiol 1960; 5:619-627.
147. Tai PC, Hayes DJ, Clark JB et al. Toxic effects of human eosinophil products on isolated rat heart cells in vitro. Biochem J 1982; 204:75-80.
148. Tart RC and van de Rijn I. Analysis of adherence of Streptococcus defectivus and endocarditis-associated streptococci to extracellular matrix. Infect Immun 1991; 59:857-862.
149. Tornebrandt K, Eskilsson J and Nobin A. Heart involvement in metastatic carcinoid disease. Clin Cardiol 1986; 9:13-19.
150. Turck JJ, Templeton CB, Bottoms GD et al. Flunixin meglumine attenuation of endotoxin-induced damage to the cardiopulmonary vascular endothelium of the pony. Am J Vet Res 1985; 46:591-596.
151. Valiathan MS, Kartha CC, Panday VK et al. A geochemical basis for endomyocardial fibrosis. Cardiovasc Res 1986; 20:679-682.
152. Van Vliet PD, Burchell HB, Titus JL. Focal myocarditis associated with pheochromocytoma. N Engl J Med 1966; 274:1102-1108.
153. Ward BJ and Donnelly JL. Hypoxia induced disruption of the cardiac endothelial glycocalyx: implications for capillary permeability. Cardiovasc Res 1993; 27:384-389.
154. Ware JA and Heistad DD. Seminars in medicine of the Beth Israel Hospital, Boston. Platelet-endothelium interactions. N Engl J Med 1993; 328:628-635.
155. Wassom DL, Loegering DA, Solley GO et al. Elevated serum levels of the eosinophil granule major basic protein in patients with eosinophilia. J Clin Invest 1981; 67:651-661.
156. Wong AJ, Pollard TD and Herman IM. Actin filament stress fibers in vascular endothelial cells in vivo. Science 1983; 219:867-869.
157. Wong MK and Gotlieb AI. In vitro reendothelialization of a single-cell wound. Role of microfilament bundles in rapid lamellipodia-mediated wound closure. Lab Invest 1984; 51:75-81.
158. Wright JP and Kirschner RH. Scanning electron microscopy of infective endocarditis. Scan Electron Microsc III 1979;
159. Young JD, Peterson CG, Venge P et al. Mechanism of membrane damage mediated by human eosinophil cationic protein. Nature 1986; 321:613-616.

160. Young WC and Herman IM. Extracellular matrix modulation of endothelial cell shape and motility following injury in vitro. J Cell Sci 1985; 73:19-32.
161. Ziegler K. Ueber die Wirkung intravenöser Adrenalininjektion auf das Gefäss-system und ihre Beziehung zur Arteriosklerose. Beitr Pathol Anat 1905; 38:229-254.

INDEX

Page numbers in italics denote figures (f) or tables (t).

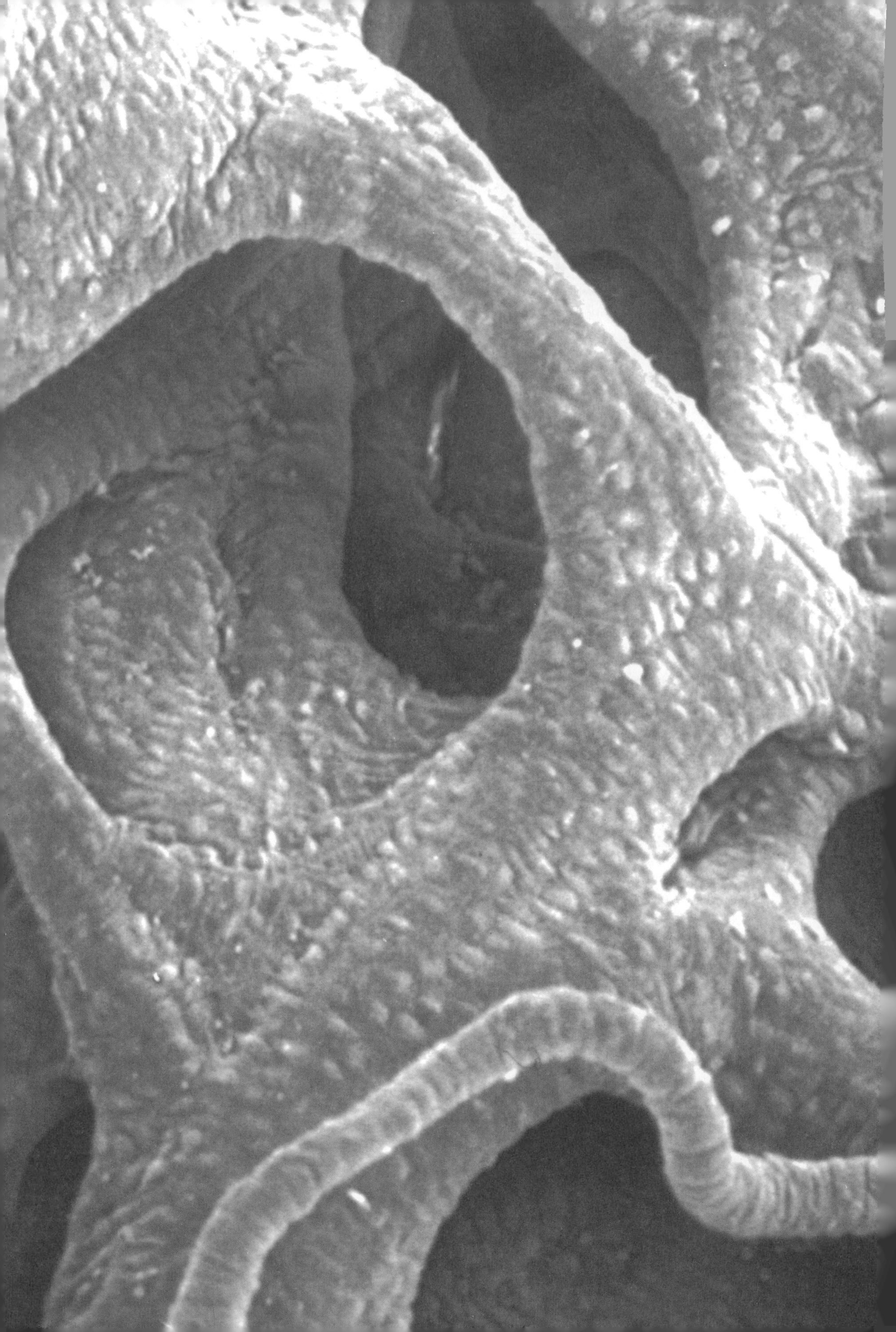